Lecture Notes in Control and Information Sciences

Volume 487

This series reports new developments in the fields of control and information sciences—quickly, informally and at a high level. The type of material considered for publication includes:

1. Preliminary drafts of monographs and advanced textbooks
2. Lectures on a new field, or presenting a new angle on a classical field
3. Research reports
4. Reports of meetings, provided they are
   (a) of exceptional interest and
   (b) devoted to a specific topic. The timeliness of subject material is very important.

Indexed by EI-Compendex, SCOPUS, Ulrich's, MathSciNet, Current Index to Statistics, Current Mathematical Publications, Mathematical Reviews, IngentaConnect, MetaPress and Springerlink.

More information about this series at http://www.springer.com/series/642

Driss Boutat · Gang Zheng

Observer Design for Nonlinear Dynamical Systems

Differential Geometric Methods

Springer

Driss Boutat
INSA Centre Val de Loire Campus
de Bourges
Bourges, France

Gang Zheng
Inria Lille-Nord Europe
Villeneuve d'Ascq, France

ISSN 0170-8643 ISSN 1610-7411 (electronic)
Lecture Notes in Control and Information Sciences
ISBN 978-3-030-73744-3 ISBN 978-3-030-73742-9 (eBook)
https://doi.org/10.1007/978-3-030-73742-9

Mathematics Subject Classification: 93-02, 53-02

This Springer imprint is published by the registered company Springer Nature Switzerland AG
The registered company address is: Gewerbestrasse 11, 6330 Cham, Switzerland

To my family

—Driss Boutat

To my wife Dan, and my children Lisa, Linda and Bona

—Gang Zheng

Preface

For engineers, there is always a need to know the states of the studied system to make important decisions to control the system or to reliably predict its future state. A simple and reliable method for obtaining system's states is to measure them directly. However, in a general complex system, it is not feasible or even possible to directly measure all states due to either technical or economic reason. An alternative to get system's states is to develop an algorithm to estimate the states. In control community, such an algorithm is named as software sensor or observer.

Observer design has attracted lots of attention since the pioneer works of Luenberger and Kalman. This is due to the numerous applications in aerospace, in robotics, in image processing, in communication, in biology and many other areas. It concerns the reconstruction of states of the studied system based on the available measurements and the model of the system (which can be uncertain). To design these observers, the concept of observability is important, and it concerns the ability of estimating states by using available data of inputs and outputs of the systems. These concepts are also relevant for diagnosis or prediction of future behavior of the system. Therefore, the observer design for nonlinear dynamical systems is an important issue in the control theory.

During last five decades, different approaches, including differential geometric and differential algebraic ones, have been successfully used to analyze the observability for dynamical systems. And many different types of observers have been proposed in the literature. However, those results related to observability analysis and observer design are dispersed in different articles and books. The students and researchers need to spend lots of time to understand how to deal with observer design problems. Therefore, a monograph dedicated to observer design for nonlinear dynamical systems would be very useful. This is exactly the main reason why this book is written.

In order to make the reading more easy, this book starts with the linear dynamical systems case, for which different equivalent definitions of observability are discussed, and several well-known observers are then presented. After having the basic ideas of observability and observer design for linear dynamical systems, we switch to investigate those two problems for nonlinear dynamical systems. When dealing with observability analysis for nonlinear dynamical systems, we adopt the differential geometric method which is one of the most used approaches. Concerning the observer design for nonlinear dynamical systems, due to the fact that the results on this topic are vast, this book just lists several well-known nonlinear observers, and then focuses on observer design technique by using differential geometric method. The idea of this technique is to transform the studied nonlinear dynamical system into a simpler form (called as observer normal form) for which we can apply existing observers to achieve the task of state estimation.

The goal of this monograph is to present how this method can be used to design observers for different types of nonlinear dynamical systems (including single/multiple outputs, fully/partially observable systems, regular/singular dynamical systems). More specifically, it will complement the existing books on observer design, by presenting some achievements in the nonlinear observer normal forms. This book contains 9 chapters, and they are organized as follows:

- Chapter 1 explains the key problem of observation for dynamical systems, and recalls the basic results on observability analysis and observer design for linear and nonlinear dynamical systems. More attention is paid for the technique of observer normal forms.
- Chapter 2 deals with a quick overview of the basic concepts, elements and theorems from differential geometry to ensure a clear understanding of the contents of this book.
- Chapter 3 discusses and analyzes single-output nonlinear dynamical systems that admit a change of coordinates via which the studied nonlinear dynamical system can be transformed into the so-called nonlinear observer normal form with output injection.
- Chapter 4 applies as well a change of coordinates to the output of nonlinear dynamical systems, compared to the result obtained in Chap. 3. It shows how to relax the geometric conditions provided in Chap. 3, by allowing nonlinear output for the desired observer normal form.
- Chapter 5 provides the new concept of the extended observer normal form by adding an auxiliary dynamics to the studied nonlinear dynamical systems. It presents as well an algorithm to compute the auxiliary dynamics as well as the change of coordinates.
- Chapter 6 relaxes the conditions presented in Chap. 3 in another way, by allowing output-depending matrix in the desired normal form, named as output-depending observer normal form.
- Chapter 7 investigates observer design problem for partially observable systems. It shows how to treat general partial observable case when applying differential geometric method, where the notion of commutativity of Lie bracket modulo a distribution is used.
- Chapter 8 extends the results on nonlinear dynamical systems with single output to the case with multiple outputs. It treats as well the partially observable case for nonlinear dynamical systems with multiple outputs.
- Chapter 9 extends the differential geometric method to design observers for nonlinear singular dynamical systems, which are governed by mixing differential

and algebraic equations. It presents the manner to regularize the singular system into a regular one, and then seek for diffeomorphism to transform the regularized system into an observer normal form with output derivative injection.

Bourges, France
Lille, France
October 2020

Driss Boutat
Gang Zheng

Contents

Chapter 1
Observability and Observer for Dynamical Systems

Abstract Before dealing with the observer design for dynamical systems, we need firstly to analyze whether the states of the studied dynamical system can be observable or not. This property is named as observability in the literature. Therefore, this chapter aims at recalling the existing results on observability analysis approaches and observer design techniques for dynamical systems. We start with the classification of observation problems, the different approaches to analyze the observability and the categories of observers. After that, we recall several important definitions and test conditions for the observability of linear dynamical systems. Some well-known observers are then summarized, including Luenberger observer, Kalman observer and so on. Finally, we present how to analyze observability for nonlinear dynamical systems, and discuss several famous nonlinear observer design techniques.

1.1 Introduction

Dynamical systems have been widely used to model different plants to be controlled in many different disciplines, ranging from biology, chemistry, to mechanics and so on. Nowadays, one of the most popular ways to model the real process is to use the state-space equations, described normally by a set of ordinary differential equations, which is named as a continuous-time system in the literature. From engineering point of view, a wide variety of information cannot directly be obtained through measurements from real systems. Due to some economical or technological reasons, we cannot place as many sensors as we want to measure the internal information, since it is expensive, or sometimes impossible. Besides, given a concrete plat, some kinds of inputs (external disturbances for example) and some parameters (such as the constants of an electrical motor, the delay in a transmission system, etc.) are unknown or are not measurable, whose estimations are sometimes needed to be used in the closed-loop controller. Similarly, more often than not, signals from sensors are

D. Boutat and G. Zheng, *Observer Design for Nonlinear Dynamical Systems*,
Lecture Notes in Control and Information Sciences 487,
https://doi.org/10.1007/978-3-030-73742-9_1

distorted and tainted by measurement noises. Therefore, although control is the final goal in the control theory, given a concrete model, in order to simulate, to control or to supervise processes, and to extract information conveyed by the signals, we often need to estimate parameters, internal variables, or the unknown inputs.

Estimation techniques are, under various guises, presented in many parts of control, signal processing, and applied mathematics. Such an important area gave rise to a huge international literature that cannot be summarized here. Roughly speaking, in automatic control, the estimation covers at least the following topics:

- Identification of uncertain parameters in the system's model, including delays;
- Estimation of state variables, which are not measured;
- Observation of the fault (unknown input) and isolation.

From control point of view, we know that the problem of parameters estimation can be converted into the problem of state observation, by regarding those parameters as additional states, and then extending the system's state, provided that they are either (piece-wise) constant, or slowly time-varying. In this book, we will focus only on the state estimation problem for the following dynamical system:

$$\begin{aligned} \dot{x} &= f(x, u) \\ y &= h(x), \end{aligned} \tag{1.1}$$

where $x \in \mathscr{X} \subset \mathbb{R}^n$, $u \in \mathscr{U} \subset \mathbb{R}^p$, and $y \in \mathscr{Y} \subset \mathbb{R}^m$ represent the state, the control input and the output of the system, respectively. It is assumed that $f : \mathscr{X} \times \mathscr{U} \to \mathscr{X}$ and $h : \mathscr{X} \to \mathscr{Y}$ are smooth. System (1.1) is said to be a single-output system if the integer $m = 1$, and to have multiple outputs if $m \geq 2$.

When dealing with state observation problem for the dynamical system, normally we need to consider the following two important issues:

- With the available information (output y, input u and their successive derivatives), is it possible to reconstruct the internal information (state x) over a time interval?
- With the available information, how to design a dynamics which enables us to reconstruct the required state x?

The first issue is related to the concepts of *Observability* and *Detectability*, and different equivalent necessary and/or sufficient conditions are proposed in the literature to judge whether the state x of system (1.1) can be reconstructed or not. The key difference between observability and detectability is the possibility to tune the estimation velocity. Precisely, note $\hat{x}(t)$ as the estimation of $x(t)$ of (1.1), then this system is said to be observable if the convergence rate of $\hat{x}(t)$ to $x(t)$ can be tuned. Otherwise, it is called to be detectable. A simple example to clarify such a difference is the following linear time-invariant dynamical system:

$$\begin{aligned} \dot{x}_1 &= x_1 + u \\ \dot{x}_2 &= -a x_2 \\ y &= x_1, \end{aligned}$$

with $a > 0$. It is clear that the first state $x_1(t)$ is observable since we can have immediately its estimation as $\hat{x}_1 = y$, while the second state $x_2(t)$ is not observable due to the fact that the dynamics of x_1 and x_2 are completely independent. In other words, we cannot reconstruct x_2 from the available information y and u. However, we notice that $x_2(t)$ converges to 0 when $t \to \infty$, since $a > 0$. This means that we can still reconstruct the state $x_2(t)$ which will converge to 0, but we cannot tune its convergence rate which in fact depends only on the system's parameter a. Therefore, the above simple system is not observable, but only detectable.

In general, observability, defining the possibility to estimate internal states of the studied system via the available measurement and its derivatives [2], has already been widely studied for different types of dynamical systems. Using the elementary algebraic method to analyze observability can be dated back to 1960s in the work of Kalman for linear dynamical systems [20]. However, the generalization of the similar theory to nonlinear dynamical systems is not so trivial. If the studied system is observable, we can then investigate how to design the estimator $\hat{x}(t)$. Generally, there exist at least two quite different categories in the literature:

- Estimation by using differentiator;
- Estimation by using observer.

If the studied system is fully observable, all its states can be then algebraically represented as a function of output/input and their derivatives [9]. This implies that an efficient differentiator, enabling to calculate the successive derivatives of the signal, is enough to estimate the states. Different types of differentiators have been proposed in the literature, such as digital differentiator [6], high-gain differentiator [8], tracking differentiator [14], algebraic differentiator [18], high-order sliding mode differentiator [29], and so on. The key issue is that the badly designed differentiator will amplify the high-frequency noise, if the signal is corrupted.

For the second category, the list of different types of observers is quite long. From the design methodology point of view, there exist two different methods. The first method tries to directly design an observer. The second method is to transform the studied system into a simpler form (called as normal form) which enables us to apply existing observers. The first idea of normal form is due to [3] for non-autonomous nonlinear dynamical systems and [25] for autonomous nonlinear dynamical systems where the authors introduced the so-called observer canonical form with output injection with all nonlinear terms being only function of the output. Then [27] gave the associated canonical form with output injection for multi-output nonlinear dynamical systems without inputs, and the result for multi-output systems with inputs was studied in [47]. Based on the above works many algorithms have been developed to generalize the existing results, including algebraic approaches [37], geometric approaches [33].

From the convergence performance point of view, we can also classify those techniques into two different categories: asymptotic and non-asymptotic.

Definition 1.1 The dynamics $\dot{\hat{x}} = \hat{f}(\hat{x}, y, u)$ with $\hat{x} \in \mathbb{R}^n$ and the user-chosen function $\hat{f}$ is an asymptotic observer of (1.1) if

$$\lim_{t\to\infty} ||\hat{x}(t) - x(t)|| = 0,$$

where $\hat{x}$ is called as the estimation of state x. It is said to be a non-asymptotic (or finite-time) observer if there exists a finite-time $T > 0$, such that

$$||\hat{x}(t) - x(t)|| = 0, \forall t \geq T.$$

Up to now, many asymptotic observers have been widely studied, such as Kalman observer [21], Luenberger observer [32], adaptive observer [23], high-gain observer [13], unknown input observer [5] and so on.

Compared to asymptotic observer, finite-time one was less studied in the literature, which, however, is well appreciated in practice. Different methods have been proposed, such as sliding mode technique [46], impulsive observer [10] and homogeneity [4]. The global finite-time observer based on homogeneity was firstly introduced by [36] for the nonlinear dynamical systems which can be transformed into a linear dynamical system with output injection. After that, [42] extended this idea and proposed a semi-global finite-time observer for the special systems with triangular structure. The global finite-time observer for such a system was studied in [34] by introducing the second gain in [43].

Generally speaking, asymptotic observers have relatively simple structures, but possess only asymptotic convergence. Sometimes, for certain reasons like safety, severe time convergence constraint is imposed for the studied system, then a non-asymptotic (convergence in a finite time) observer is desired. In this book, we will briefly investigate several types of asymptotic and non-asymptotic observers in each chapter.

It is worth emphasizing again that the observability and observer design are not two equivalent concepts. A system is observable implies the existence of certain observers, but the inverse is not always true, since the system can be detectable. In other words, the deduced observability condition is only sufficient for certain types of observers, and other sorts of observers (Luenberger-like, high-gain,...) might ask for additional conditions, for example, the well-known Lipschitz condition when treating nonlinear dynamical systems.

In this book, we restrict our study only on the observability and will not discuss the detectability. In order to avoid the ambiguity, in this book, when we discuss the observer design, if we do not explicitly mention the words *detectable/detectability*, then *observer* always implies the *tunable* asymptotic or non-asymptotic observer.

The following sections will give a brief overview of observability conditions and observer design techniques which we can find in the literature. We will firstly present those results for linear time-invariant dynamical systems, since it is the simplest case, and some results are equivalent for nonlinear dynamical systems. Then we will clarify how those results for linear dynamical systems are extended to treat nonlinear dynamical systems.

1.2 Observability and Observer for Linear Dynamical Systems

Linear time-invariant dynamical systems are commonly written in the following state-space equation:

$$\begin{aligned} \dot{x} &= Ax + Bu \\ y &= Cx, \end{aligned} \tag{1.2}$$

where $x \in \mathbb{R}^n$, $u \in \mathbb{R}^p$ and $y \in \mathbb{R}^m$ represent the state, the control input and the measurement, respectively, with $A \in \mathbb{R}^{n\times n}$, $B \in \mathbb{R}^{n\times p}$ and $C \in \mathbb{R}^{m\times n}$ being constant matrices.

During last five decades, many different methods have been proposed for observability analysis and observer design for (1.2). Due to the fact that the studied system is linear, we can normally use the elementary algebra (vector space, linear transformation, rank, image) to analyze its properties, such as observability and detectability of systems with known or unknown inputs [35]. For such systems, several different but equivalent definitions of observability have been proposed in the literature [45]. The most common way is the distinguishability, i.e., different initial conditions yield different outputs. Since zero initial conditions produce zero outputs, the above property is equivalent to say that non-zero initial conditions imply non-zero outputs. Moreover, since the output of a system is uniquely determined by its initial conditions, we can also characterize the observability for linear dynamical systems as the ability to reconstruct initial conditions. For linear dynamical systems, this expression leads to the well-known invertibility of the observability Gramian, which can be used as a criteria to test whether the studied system is observable or not. Of course, some other checkable conditions have been proposed to test the observability, such as the famous Kalman rank condition [20], Popov–Belevich–Hautus (PBH) test condition [1], unobservable space and invariant zero set [15]. Some of those mentioned results will be recalled in the following.

1.2.1 Observability Analysis

Since the objective is to estimate $x(t)$ from available information $y(t)$ and $u(t)$, it is natural to consider firstly the output equation as follows:

$$y = Cx.$$

The trivial case is that the matrix C is left invertible, i.e., $m \geq n$ and rank $C = n$. In such a case, the state can be instantaneously obtained via

$$x(t) = \left[C^T C\right]^{-1} C^T y(t).$$

This trivial case is, however, not realizable in most of applications, since it explicitly implies that

- the number of measurement sensors should be large enough (i.e., $m \geq n$);
- the placement of those sensors should be intelligent enough to satisfy rank $C = n$.

If one of those two conditions is not valid, then instantaneous estimation of $x(t)$ only from $y(t)$ is not possible. We should then rely on the information provided by the dynamical equation $\dot{x} = Ax + Bu$ to check whether it is possible to estimate $x(t)$. Due to this reason, the estimation of $x(t)$ will not any more instantaneous, but over a time interval.

Precisely, for the simple linear ODE described in (1.2), its analytic solution can be written as

$$x(t) = e^{A(t-t_0)}x(t_0) + \int_{t_0}^{t} e^{A(t-s)}Bu(s)ds, \tag{1.3}$$

where t_0 is the initial time, $x(t_0)$ is the corresponding unknown initial condition and $e^{A(t-t_0)}$ is named as *state transition matrix* in the literature. Since the input u and the constant matrices A and B are all known, the intuitive definition of observability can be stated as follows.

Definition 1.2 Observability of the state $x(t)$ is equivalent to the reconstructibility of the initial condition $x(t_0)$ over the time interval $[t_0, t]$.

The reason why the reconstruction of $x(t_0)$ is over the time interval $[t_0, t]$ is that (1.3) contains the value t_0 and the integration of $u(t)$ from t_0 to t. Consequently, the observability problem is to deduce the conditions under which $x(t_0)$ can be estimated through $y(s)$ and $u(s)$ for $s \in [t_0, t]$. By linking the measurable output to the initial condition $x(t_0)$ we have

$$y(t) = Ce^{A(t-t_0)}x(t_0) + \int_{t_0}^{t} Ce^{A(t-s)}Bu(s)ds. \tag{1.4}$$

Remark 1.1 We can observe from (1.4) that the term $\int_{t_0}^{t} Ce^{A(t-s)}Bu(s)ds$ has no real contribution to the observability condition since the matrices A, B, C and the control $u(s)$ for $s \in [t_0, t]$ are known. In fact, we can merely redefine a virtual measurement as

$$\bar{y}(t) = y(t) - \int_{t_0}^{t} Ce^{A(t-s)}Bu(s)ds, \tag{1.5}$$

which leads to

$$\bar{y}(t) = Ce^{A(t-t_0)}x(t_0). \tag{1.6}$$

The observability conditions should be equivalent for both (1.4) and (1.6). Due to this observation and for the sake of simplicity, we can set $B = 0$ in (1.4).

According to Definition 1.2, the observability of a linear time-invariant dynamical system refers to estimate its initial condition $x(t_0)$ via available information y and u. Note that the input–output equation defined in (1.4) reveals the relations between $x(t_0)$ and y, u, based on which several different but equivalent definitions have been stated in the literature. For the sake of simplicity, denote $x_0 = x(t_0)$ as the initial condition of (1.2), $x(t, x_0, u)$ as its solution defined in (1.3) with the initial condition x_0 and the input u, and $y(t, x_0, u)$ as its associated output defined in (1.4). From (1.4), we observe that it defines a one-to-one mapping between $y(t)$ and x_0. Given two different initial conditions x_0 and $\bar{x}_0$, a straightforward calculation shows that

$$y(t, x_0, u) - y(t, \bar{x}_0, u) = Ce^{A(t-t_0)}\left[x_0 - \bar{x}_0\right], \forall u, \forall t \geq 0. \tag{1.7}$$

Following the above equation, different interpretations lead to different definitions of observability. One interpretation is that different initial conditions should yield different outputs, which gives the following definition.

Definition 1.3 System (1.2) is said to be observable if for any initial conditions $x_0 \in \mathbb{R}^n$ and $\bar{x}_0 \in \mathbb{R}^n$ with $x_0 \neq \bar{x}_0$, we have

$$y(t, x_0, u) \neq y(t, \bar{x}_0, u), \forall u, \forall t \geq 0.$$

The above statement enables us to define the concept of indistinguishability of the state which can produce the same output via different states.

Definition 1.4 Two states $x_0 \in \mathbb{R}^n$ and $\bar{x}_0 \in \mathbb{R}^n$ of system (1.2) are called indistinguishable if

$$y(t, x_0, u) = y(t, \bar{x}_0, u), \forall u, \forall t \geq 0.$$

Based on Definition 1.4 we can then have another way to define the observability.

Definition 1.5 System (1.2) is said to be observable if it does not admit any indistinguishable state.

We have reviewed some different but equivalent definitions of observability for linear time-invariant systems, which, however, are just conceptual, and do not provide a feasible way to check directly the observability of linear time-invariant dynamical system (1.2). Therefore, several checkable methods have been proposed in the literature to judge the observability.

The first method is related to identification technique. Since the objective of observability is to estimate the state $x(t)$, it can be viewed as the following constrained optimization problem:

$$\hat{x} = \arg\min_{x\in\mathbb{R}^n} ||y - Cx||^2$$
$$\text{subject to:} \quad \dot{x} = Ax + Bu.$$

Since solving directly the above constrained minimization problem is not trivial, we can remove this constraint by replacing its analytic solution

$$\hat{x}(t_0) = \arg\min_{x(t_0)\in\mathbb{R}^n} ||\bar{y} - Ce^{A(t-t_0)}x(t_0)||^2, \tag{1.8}$$

where $\bar{y}$ was defined in (1.5). Define the following matrix:

$$\mathscr{W}_{O,T} = \int_{t_0}^{t_0+T} \left[Ce^{A(s-t_0)}\right]^T Ce^{A(s-t_0)} ds, \tag{1.9}$$

where T is positive. This matrix $\mathscr{W}_{O,T}$ is conventionally named as Observability Gramian. Since the initial condition $x(t_0)$ is constant, according to the theory of parameter identification [31], the condition of the existence of the solution for the minimization problem (1.8) is that there exists a positive T such that the observability Gramian $\mathscr{W}_{O,T}$ is invertible.

Theorem 1.1 *System (1.2) is observable if and only if there exists $T > 0$ such that*

$$rank\ \mathscr{W}_{O,T} = n. \tag{1.10}$$

Condition (1.10) is checkable, but not so direct since it depends on the choice of positive value T. In practice, when $T \to \infty$ and A is Hurwitz, the observability Gramian is the solution of the following algebraic equation

$$A^T \mathscr{W}_{O,\infty} + \mathscr{W}_{O,\infty} A + C^T C = 0,$$

which can be easily computed.

Another more direct way to judge the observability is the well-known Kalman's observability rank condition. Define the following matrix:

$$d\mathscr{O} = \begin{bmatrix} C \\ CA \\ \vdots \\ CA^{n-1} \end{bmatrix}, \tag{1.11}$$

which is named as *observability matrix*. It can be noticed that the observability matrix is calculated up to order $n-1$, and this is due to Cayley–Hamilton Theorem [41] which states that CA^k for $k \geq n$ are linearly dependent of $\{C, CA, \ldots, CA^{n-1}\}$. Based on the defined observability matrix, we have the following result.

Theorem 1.2 *System (1.2) is observable if and only if*

$$rank\ d\mathscr{O} = n. \tag{1.12}$$

Remark 1.2 A simple way to interpret the above condition is as follows. Consider system (1.2) with $B = 0$, since we know the output $y(t)$, thus its high-order derivatives are known as well. Hence we have

$$\begin{bmatrix} y(t_0) \\ \dot{y}(t_0) \\ \vdots \\ y^{(n-1)}(t_0) \end{bmatrix} = \begin{bmatrix} C \\ CA \\ \vdots \\ CA^{n-1} \end{bmatrix} x(t_0),$$

where $y^{(i)}$ implies the ith derivative of y with respect to t. Therefore, if the observability matrix $d\mathcal{O}$ is non-singular, then the initial condition $x(t_0)$ can be reconstructed. As we will see hereafter, the same idea has been extended to deduce observability conditions for nonlinear dynamical systems.

Compared to the observability Gramian $\mathcal{W}_{O,T}$, it is much easier to calculate the observability matrix $d\mathcal{O}$. Another easily checkable condition is the PBH test [17] which is stated as follows.

Theorem 1.3 *System (1.2) is observable if and only if*

$$rank \begin{bmatrix} \lambda I - A \\ C \end{bmatrix} = n, \forall \lambda \in \mathbb{C}. \tag{1.13}$$

Remark 1.3 The conditions (1.10), (1.12) and (1.13) are all equivalent necessary and sufficient conditions to check the observability. Although (1.13) is less checkable than (1.12), it provides another advantage: check the detectability. Notice that λ can be seen as the eigenvalues of A. Since negative eigenvalues mean the corresponding dynamical parts are stable (i.e., asymptotically converge to zero), the detectability requires only that the condition (1.13) is valid for all positive eigenvalues, i.e.,

$$\text{rank} \begin{bmatrix} \lambda I - A \\ C \end{bmatrix} = n, \forall \lambda \in \mathbb{C}^+.$$

Example 1.1 Consider the following linear time-invariant dynamical system:

$$\begin{aligned} \dot{x}_1 &= -x_1 - x_2 \\ \dot{x}_2 &= x_1 - x_3 \\ \dot{x}_3 &= -x_3 \\ y &= x_2 - x_3. \end{aligned}$$

The state matrix is given by

$$A = \begin{bmatrix} -1 & -1 & 0 \\ 1 & 0 & -1 \\ 0 & 0 & -1 \end{bmatrix} \quad \text{and} \quad C = \begin{bmatrix} 0 & 1 & -1 \end{bmatrix}.$$

By solving the following algebraic equation:

$$A^T \mathcal{W}_{O,\infty} + \mathcal{W}_{O,\infty} A + C^T C = 0,$$

we have

$$\mathscr{W}_{O,\infty} = \begin{bmatrix} 1/2 & 1/2 & -5/6 \\ 1/2 & 1 & -7/6 \\ -5/6 & -7/6 & 5/3 \end{bmatrix},$$

which yields

$$\text{rank}\mathscr{W}_{O,\infty} = 3.$$

Thus, the observability Gramian is invertible and the studied system is observable.

We can also test PBH condition to judge the observability. For this, we calculate

$$\text{rank}\begin{bmatrix} \lambda I - A \\ C \end{bmatrix} = \text{rank}\begin{bmatrix} \lambda+1 & 1 & 0 \\ -1 & \lambda & 1 \\ 0 & 0 & \lambda+1 \\ 0 & 1 & -1 \end{bmatrix} = 3, \forall \lambda \in \mathbb{C}$$

and it implies as well the studied system is observable.

The simplest manner to judge the observability is no doubt the Kalman rank condition. It is easy to calculate that

$$CA = \begin{bmatrix} 1 \, 0 \, 0 \end{bmatrix}, \quad CA^2 = \begin{bmatrix} -1 \, -1 \, 0 \end{bmatrix},$$

thus we have

$$\text{rank}\begin{bmatrix} C \\ CA \\ CA^2 \end{bmatrix} = \text{rank}\begin{bmatrix} 0 & 1 & -1 \\ 1 & 0 & 0 \\ -1 & -1 & 0 \end{bmatrix} = 3.$$

Therefore, the Kalman rank condition (1.12) is satisfied and the studied system is observable. □

1.2.2 Observer Design

Consider the following linear time-invariant dynamical system

$$\begin{aligned} \dot{x} &= Ax + Bu \\ y &= Cx, \end{aligned} \tag{1.14}$$

where $x \in \mathbb{R}^n$, $u \in \mathbb{R}^p$ and $y \in \mathbb{R}^m$. Assume that it is observable, which is equivalent to suppose that the condition rank $d\mathscr{O} = n$, with $d\mathscr{O}$ being defined in (1.11), is satisfied which in fact depends only on the matrices A and C. Thus, when system (1.14) is observable, for simplicity we say that the pair (A, C) is observable.

Luenberger Observer

The first type of asymptotic observers for (1.14) is the so-called Luenberger observer [32], whose dynamics is given as follows:

$$\dot{\hat{x}} = A\hat{x} + Bu + K(y - C\hat{x}),$$

where $\hat{x} \in \mathbb{R}^n$ represents the estimation of state, K is a gain matrix to be chosen. If (A, C) is observable, then there always exists a matrix K such that $(A - KC)$ is Hurwitz, i.e., all eigenvalues of $(A - KC)$ are located in the left half of the complex plane.

Denote the estimation error by $e(t) = x(t) - \hat{x}(t)$, then we obtain the following observation error dynamics

$$\dot{e} = (A - KC)e.$$

Due to the fact that the chosen K makes $(A - KC)$ be Hurwitz, then we can conclude that $e(t)$ tends to 0 as $t \to +\infty$. In other words, $\hat{x}(t)$ asymptotically converges to $x(t)$.

It is worth noting that the structure of Luenberger observer is quite simple, but the choice of the gain matrix K is not unique. Large K will place the eigenvalues of $(A - KC)$ more negative (their real parts), thus the estimation error e will converge to 0 faster. However, to choose such a large K is not a good solution when the output is corrupted by measurement noise. In this case, $e(t)$ will not any more converge to 0, but into a neighborhood of 0. The large K will then amplify this neighborhood.

Kalman–Bucy Filter

Consider again system (1.14) with noise both in the model and in the output as follows:

$$\begin{aligned} \dot{x} &= Ax + Bu + w_x \\ y &= Cx + w_y, \end{aligned} \tag{1.15}$$

where $w_x \in \mathbb{R}^n$ and $w_y \in \mathbb{R}^m$ denote the noises and assumed to be Gaussian.

Obviously, we can still design a Luenberger observer for such a system. In this case, it is expected that the estimation error e will not converge to 0. However, an optimal Luenberger observer should have a gain matrix K such that the square error $e^T e$ is minimal even both the model and the output are corrupted by noise. Such an optimal observer is named as Kalman–Bucy observer [21], which is of the following form:

$$\begin{aligned} \dot{\hat{x}} &= A\hat{x} + Bu + K(y - C\hat{x}) \\ K &= PC^T R^{-1} \\ \dot{P} &= AP + PA^T + Q - PC^T R^{-1} CP, \end{aligned} \tag{1.16}$$

where R and Q are the covariance of the noise w_y and w_x, and P represents the covariance of the estimation error $e(t)$. It has been proven in many textbooks that such a choice of K is optimal in the sense of minimizing $e^T e$ when the added model and output noises are Gaussian.

Observer Canonical Form

If system (1.14) is observable, i.e., the pair (A, C) fulfills the observability rank condition, then there always exists a change of coordinates $z = Tx$ such that it can be transformed into the following observer canonical form:

$$\dot{z} = A_O z + \beta y + \bar{B}u \quad (1.17)$$

$$y = C_O z, \quad (1.18)$$

where

$$A_O = \begin{pmatrix} 0 & \cdots & 0 & 0 & 0 \\ 1 & \cdots & 0 & 0 & 0 \\ \vdots & \ddots & \ddots & \cdots & \vdots \\ 0 & \cdots & 1 & 0 & 0 \\ 0 & \cdots & 0 & 1 & 0 \end{pmatrix}, C_O = \begin{pmatrix} 0 & \cdots & 0 & 1 \end{pmatrix}, \quad (1.19)$$

which is named as the Brunovsky form [45].

Such a transformation can be achieved by applying the change of coordinates $z = Tx$ with the rows of T are determined by the following equations:

$$T_n = C \quad (1.20)$$

$$T_{n-k} = CA^k + \sum_{i=1}^{k} p_{n-i} CA^{k-i}, \quad \text{for } 1 \leq k \leq n-1, \quad (1.21)$$

where p_i for $0 \leq i \leq n-1$ are coefficients of the characteristic polynomial of matrix A, i.e.,

$$P_A(s) = p_0 + p_1 s + \cdots + p_{n-1}s^{n-1} + s^n$$

and

$$\bar{B} = TB$$

and $\beta = [\beta_1, \ldots, \beta_n]^T$ is vector field of the following form:

$$\beta_n = -p_{n-1}$$

$$\beta_{n-k} = -p_{n-(k+1)}, \quad \text{for } 1 \leq k \leq n-1.$$

For the observable canonical form (1.18), since y and u are both known, it is easy to design a Luenberger observer of the following form:

$$\dot{\hat{z}} = A_O\hat{z} + \beta y + \bar{B}u + K(y - C_O\hat{z}),$$

which gives the following observation error dynamics:

$$\dot{e} = (A_O - KC_O)e,$$

where $e = z - \hat{z}$. Due to the special form of A_O and C_O, it is easy to find K such that $(A_O - KC_O)$ is Hurwitz. Once e tends to zero, since T is an invertible mapping, we can conclude that $T^{-1}\hat{z}$ is the estimation of x.

1.3 Observability and Observer for Nonlinear Dynamical Systems

Nonlinear dynamical systems are commonly written in the following state-space equation:

$$\begin{aligned} \dot{x} &= f(x, u) \\ y &= h(x), \end{aligned} \tag{1.22}$$

where $x \in \mathscr{X} \subset \mathbb{R}^n$, $u \in \mathscr{U} \subset \mathbb{R}^p$ and $y \in \mathbb{R}^m$ represent the state, the control input and the measurement, respectively, with $\mathscr{X}$ and $\mathscr{U}$ being the neighborhood of 0 with appropriate dimension. In this book, it is assumed that f and h are smooth, i.e., $f \in C^\infty$ and $h \in C^\infty$.

Methodologically, two different approaches are normally used in the literature to study observability for nonlinear dynamical systems

- Differential algebraic method: regard the differential as an operator, and analyze the nonlinear dynamical systems by using the theorem of module (differential ring, differential field...) [7];
- Differential geometric method: consider the nonlinear dynamical system evolving on certain manifolds, and analyze the dynamics by using the concepts of differential geometry (vector fields, Lie derivative, Lie bracket...) [19].

For nonlinear dynamical systems, till 1970s, Hermann and Krener showed in [16] that the differential geometric method was useful to analyze its properties, including observability. This method opened a door in the control domain to study many control problems for nonlinear dynamical systems, and lots of results are published by using differential geometric method. It was until the middle of 1980s that the differential algebraic method was introduced by Fliess [12] to analyze nonlinear dynamical systems. In the following, we will recall some results on observability analysis for nonlinear dynamical systems by using differential geometric method.

1.3.1 Observability Analysis

Similar to the case of linear time-invariant dynamical system, the observability analysis for nonlinear dynamical system is also to estimate $x(t)$ from available information $y(t)$ and $u(t)$. Generally speaking, we are seeking an injective mapping from y, u and their derivatives to x. Similar to the observability analysis for linear time-invariant dynamical systems, we can firstly consider the output equation

$$y = h(x).$$

The trivial case is that the function h defines an into mapping. It is worth noting that, due to the different nonlinear properties of the function h, this mapping might be injective globally or locally, which is the key difference compared to the linear case. To clarify this point, consider the following two different functions

$$y = h(x) = e^x$$

and

$$y = h(x) = \sin x.$$

It is clear that $y = e^x$ defines a global (for $x \in \mathbb{R}$) one-to-one mapping, while $y = \sin x$ is only a local one for $x \in [(2k-1)\pi, (2k+1)\pi)$. Given a general nonlinear function h (i.e., without specifying its nonlinearity), normally we can only characterize its local property, and this is the main reason why we consider in the most cases the local observability when studying nonlinear dynamical systems.

For the trivial case where h has already defined an injective mapping, it is not necessary to consider the dynamical equation $\dot{x} = f(x, u)$. The sufficient condition to judge the observability can be stated as: System (1.22) is locally observable if

$$\text{rank} \left[\frac{\partial h}{\partial x} \right] = n, \forall x \in \mathscr{X}.$$

Note that if the above condition is satisfied, then according to Implicit Function Theorem, there exists a local inverse mapping that gives x as a unique function of y, i.e., x can be uniquely determined by y. In this sense, x is locally observable. It is worth noting that this condition implicitly requires that $h(x)$ has n independent components.

Due to the same reasons (economic or technique) for linear time-invariant dynamical systems, this trivial case is hard to be satisfied. Thus, it is not possible to use only the instantaneous value of y to achieve estimation of x. In this case, we might consider the dynamics $\dot{x} = f(x, u)$ to estimate $x(t)$ by using both the values of y and u over a time interval.

As we have seen in Sect. 1.2.1 that the observability of the state $x(t)$ for linear time-invariant dynamical system (1.2) is equivalent to the reconstructibility of its initial condition $x(t_0)$, this statement is also valid for nonlinear dynamical systems. We would like to remark that the observability does not require the uniqueness of solutions, which can be illustrated by the following simple example

$$\begin{aligned} \dot{x} &= f(x, u) \\ y &= x, \end{aligned}$$

with $x \in \mathbb{R}^n$. It is evident that the above system is always observable even if it admits several different solutions.

For the nonlinear dynamical system (1.22), denote $x_0 = x(t_0)$ as its initial condition, $x(t, x_0, u)$ as its solution with the initial condition x_0 and the input u, and $y(t, x_0, u)$ as its associated output. We can then define observability via the indistinguishability for nonlinear dynamical systems, as Definitions 1.4 and 1.5 for linear time-invariant dynamical systems. However, unlike linear case, the observability for nonlinear case might be global or local, and this leads to the following modified definitions.

Definition 1.6 Two states $x_0 \in \mathbb{R}^n$ and $\bar{x}_0 \in \mathbb{R}^n$ of system (1.22) are called indistinguishable if

$$y(t, x_0, u) = y(t, \bar{x}_0, u), \forall u, \forall t \geq 0.$$

They are $\mathscr{X}$-indistinguishable if the above condition is satisfied for $x_0 \in \mathscr{X} \subset \mathbb{R}^n$ and $\bar{x}_0 \in \mathscr{X} \subset \mathbb{R}^n$.

Based on Definition 1.6 we can then have different observability definitions for nonlinear dynamical systems [16].

Definition 1.7 System (1.22) is said to be globally observable if there is no indistinguishable state in $\mathbb{R}^n$.

Definition 1.8 System (1.22) is said to be weakly observable at x if there is a neighborhood Ω of x such that for every $\bar{x} \in \Omega$, x and $\bar{x}$ are distinguishable. System (1.22) is said to be weakly observable if it is so at every $x \in \mathscr{X}$.

Definition 1.9 System (1.22) is said to be locally weakly observable at x if there is a neighborhood Ω of x such that for every neighborhood V contained in Ω of x and every $\bar{x} \in V$, x and $\bar{x}$ are distinguishable. System (1.22) is said to be locally weakly observable if it is so at every $x \in \mathscr{X}$.

Remark 1.4 Global observability means that x can be distinguished from any other states in $\mathbb{R}^n$, where weak observability implies that x can distinguish any state in its neighborhood $\mathscr{X}$. Local weak observability requires that x is weakly observable in its neighborhood $V \subset \mathscr{X}$ without leaving V. Obviously, the concept of local weak observability is the strongest definition, but it is the most interesting one in practice.

The above different concepts of observability are due to the nonlinear property of dynamical systems. Another characteristic compared to linear time-invariant dynamical systems is the dependence of the input u. In Sect. 1.2.1, we have shown that the input u has no influence to deduce the observability conditions for linear time-invariant dynamical systems, and here we will show that it is not true for nonlinear dynamical systems. Consider the following dynamical system:

$$\begin{aligned}\dot{x}_1 &= x_1 + \alpha(u)x_2 \\ \dot{x}_2 &= x_2 x_1 \\ y &= x_1,\end{aligned}$$

where $\alpha(u)$ is a function of the input u. It is easy to obtain that, if $\alpha(u) \neq 0$ for all $u \in \mathscr{U}$, then the above system is globally observable since $x_1 = y$ and $x_2 = \frac{1}{\alpha(u)}[\dot{y} - y]$. Unfortunately, if $\alpha(u) = 0$ for some input u, then the above system is not observable. In other words, the observability of nonlinear dynamical systems might depend on the value of input, and we call the inputs (controls) that distinguish all pairs of states as universal inputs. Otherwise, we name them as singular inputs.

Definition 1.10 The dynamical system (1.22) is uniformly observable if all inputs are universal.

For linear time-invariant dynamical systems, we can obtain analytic solution of the input–output mapping, and this enables us to use the elementary algebraic approach to test the observability, such as the observability Gramian (see Theorem 1.1), the Kalman rank condition (see Theorem 1.2) and the PBH test (see Theorem 1.3). However, to deduce the observability condition from the indistinguishability for nonlinear dynamical systems is a challenging task since normally the expression of the input–output mapping cannot be obtained analytically. As indicated by the title of this book, we adopt the differential geometric method to characterize the observability condition for nonlinear dynamical systems. For this, let us introduce the following $\mathbb{R}$-vector space:

$$\mathscr{O} = \text{span}_{\mathbb{R}}\{L_{f_{u_k}} L_{f_{u_{k-1}}} \ldots L_{f_{u_1}} h_i, \quad \text{for } 1 \leq i \leq m \text{ and } u_1, \ldots, u_k \in \mathscr{U}\},$$

which is called the observation space, where $f_{u_i} = f(x, u_i)$ and $L_{f_{u_1}} h_i$ is the Lie derivative of h_i in the direction of f_{u_1}, i.e., $L_{f_{u_1}} h_i = \frac{\partial h_i}{\partial x} f_{u_1}$, which will be detailed in Chap. 2.

Define $d\mathscr{O}$ as the $\mathbb{R}$-vector space of the differentials of the observation space $\mathscr{O}$

$$d\mathscr{O} = \text{span}_{\mathbb{R}}\{dl \ \ with \ \ l \in \mathscr{O}\}, \tag{1.23}$$

then we have the following definition of observability rank condition.

Definition 1.11 System (1.22) is said to satisfy observability rank condition at x if

$$\dim(d\mathscr{O}|_x) = n. \tag{1.24}$$

It is said to satisfy observability rank condition in $\mathscr{X}$ if this is true for every $x \in \mathscr{X}$.

Using the observability rank condition, we can then have the following theorem.

Theorem 1.4 *System (1.22) is locally weakly observable at x if the observability rank condition is satisfied at x. System (1.22) is locally weakly observable if it satisfies the observability condition for every $x \in \mathscr{X}$.*

The above theorem presents a constructive way to check the observability for nonlinear dynamical system (1.22). For linear time-invariant dynamical systems, condition (1.24) is equivalent to the Kalman rank condition. However, it is too general to be used directly due to the following two reasons:

- Firstly, unlike linear case where Cayley–Hamilton Theorem tells us to stop at the order $n-1$ when checking the rank condition, there does not exist a fixed value k for nonlinear case in general;
- Secondly, as we have discussed before, the observability rank condition depends on the inputs.

In practice, a less general case is to consider only nonlinear dynamical system (1.22) without inputs. The reason is that the feedback controller u in (1.22) might be a smooth function of the state x, and this leads to investigate the observability for the following dynamical system:

$$\begin{aligned} \dot{x} &= f(x) \\ y &= h(x), \end{aligned} \tag{1.25}$$

where f and h are both assumed to be smooth. Similar to the arguments in Remark 1.2 for linear time-invariant dynamical systems, we can define a mapping between the output and the state by calculating high-order derivatives of y, which gives

$$\begin{bmatrix} y(t) \\ \dot{y}(t) \\ \vdots \\ y^{(k)}(t) \end{bmatrix} = \begin{bmatrix} h(x) \\ L_f h(x) \\ \vdots \\ L_f^k h(x) \end{bmatrix} = \mathcal{O}(x),$$

where $L_f^k h$ are the so-called kth Lie derivative of h in the direction of the vector field f which will be detailed in the next chapter, $y^{(k)}$ represents the kth time derivative of y and $\mathcal{O}(x)$ is called observability mapping. For a certain integer $k > 0$, if $\mathcal{O}(x)$ is a local diffeomorphism, then we can reconstruct the state x from the output and its derivatives.

Theorem 1.5 *System (1.25) is locally observable at x if for a certain k such that*

$$\dim d\mathcal{O} = rank \begin{bmatrix} dh(x) \\ dL_f h(x) \\ \vdots \\ dL_f^k h(x) \end{bmatrix} = n. \tag{1.26}$$

It is locally (or globally) observable if the above rank condition is true for every $x \in \mathcal{X}$ (or for every $x \in \mathbb{R}^n$).

We would like to emphasize that the above rank condition is only sufficient. Also, we have stated before that it is not possible to fix the value k for nonlinear dynamical systems, while $k = n - 1$ for linear case due to Cayley–Hamilton Theorem. This can be shown by the following example.

Example 1.2 Consider the following dynamical system:

$$\begin{aligned} \dot{x} &= 0 \\ y &= h(x) = x^3. \end{aligned}$$

It is easy to see that $d\mathcal{O}$ is spanned by $3x^2 dx$. By verifying the rank condition (1.26) for $k = 1$, we can only state that the example is locally observable since $\dim d\mathcal{O} = 1$ for $x \neq 0$. However, we cannot conclude that the example is not observable at $x = 0$ even if $\dim d\mathcal{O} = 0$ when $x = 0$. In fact, this example is obviously globally observable since $x \to x^3$ is a bijective mapping.

By calculating the high-order derivatives of y, we have

$$\begin{bmatrix} dh(x) \\ dL_f h(x) \\ \vdots \\ dL_f^k h(x) \end{bmatrix} = \begin{bmatrix} 3x^2 \\ 6x \\ 6 \end{bmatrix},$$

whose rank is always equal to 1 for all $x \in \mathbb{R}$. Therefore, the studied example is globally observable, and we found $k = 3$ when judging its observability property. □

1.3.2 *Observer Design*

Consider the following nonlinear dynamical system:

$$\begin{aligned} \dot{x} &= f(x) \\ y &= h(x), \end{aligned} \tag{1.27}$$

where $x \in \mathcal{X} \subset \mathbb{R}^n$, $u \in \mathcal{U} \subset \mathbb{R}^p$, $y \in \mathbb{R}^m$ and the functions f and h are smooth, i.e., $f \in C^\infty$ and $h \in C^\infty$. It is assumed that system (1.27) is locally observable in the sense that the observability rank condition (1.26) is satisfied for all $x \in \mathcal{X}$.

Since the objective is to reconstruct the state $x(t)$, the estimation can be regarded as the following nonlinear constrained optimization problem

$$\begin{aligned} \hat{x} &= \arg\min_{x \in \mathcal{X}} ||y - h(x)||^2 \\ \text{subject to:} &\quad \dot{x} = f(x). \end{aligned}$$

Note that for linear case where $f(x) = Ax$ and $h(x) = Cx$, since its analytic solution is known, the above constrained optimization problem can be converted to the unconstrained minimization problem (1.8) where classical least-square method can be applied to identify the initial condition $x(t_0)$, yielding the condition on the observability Gramian. However, for nonlinear case, it is not anymore possible since it is normally difficult or impossible to find the analytic solution to remove the dynamical constraint. We can only use constrained optimization techniques to numerically solve this minimization problem. This book will not discuss such a method, and interested readers can refer to related literature for more details.

Similar to linear case, we can also use certain dynamical systems to achieve the estimation of $x(t)$ for (1.27). The following will outline some well-known nonlinear observers.

Nonlinear Luenberger Observer

For the considered nonlinear dynamical system (1.27), the corresponding nonlinear Luenberger observer is of the following form:

$$\dot{\hat{x}} = f(\hat{x}) + \left[\frac{\partial T(\hat{x})}{\partial \hat{x}}\right]^{-1} \left[K(y) - K(h(\hat{x}))\right],$$

where $T(.)$ and $K(.)$ are two functions of their arguments with appropriate dimension to be determined and $T(.)$ is a diffeomorphism, which are the solutions of following partial differential equation:

$$\frac{\partial T(x)}{\partial x} f(x) = HT(x) + K(h(x)), \tag{1.28}$$

where H is constant matrix which should be chosen as Hurwitz one. $\left[\frac{\partial T(\hat{x})}{\partial \hat{x}}\right]^{-1}$ implies the inverse of the Jacobian of the diffeomorphism $T(x)$.

Denote $z = T(x)$ and $\hat{z} = T(\hat{x})$, then we have

$$\dot{z} = \frac{\partial T(x)}{\partial x} f(x) = Hz + K(y)$$

and

$$\begin{aligned}\dot{\hat{z}} &= \frac{\partial T(\hat{x})}{\partial \hat{x}} \dot{\hat{x}} = \frac{\partial T(\hat{x})}{\partial \hat{x}} f(\hat{x}) + K(y) - K(h(\hat{x})),\\ &= H\hat{z} + K(y)\end{aligned}$$

which yields the following observation error dynamics

$$\dot{e} = He,$$

where $e = z - \hat{z}$. Since H is chosen to be Hurwitz, e tends to zero asymptotically. Moreover, if the solution $T(x)$ of the PDE (1.28) is a diffeomorphism, then we can concluded that $\hat{x}$ tends to x asymptotically.

The key issue of this method is to solve the PDE (1.28). Note that, for linear time-invariant dynamical systems, the PDE becomes the following matrix equation:

$$TA = HT + KC,$$

thus, if (A, C) is observable, then we can choose $T = I$ which yields

$$H = A - KC.$$

By choosing K properly, we can always ensure that H is Hurwitz.

For nonlinear systems, given f and h, the solution $T(x)$ of this PDE always exists, but might not be injective, thus some sufficient conditions should be imposed for (1.27) (see [28]). Note that the dimension of $T(x)$ might be equal or bigger than that of x. For the case where z and x have the same dimension, [22] gave sufficient conditions for the local solvability of the PDE (1.28). The solution is obtained by linearizing f and h around the equilibrium point 0. Due to this reason, the solution is local, so is for the proposed nonlinear Luenberger observer. However, this result is also valid for nonlinear analytic systems (not necessary to be Lipschitz). Some extensions have been done in [24] for Lipschitz continuous systems.

Extended Kalman Filter

Another observer for general system (1.27) is the so-called extended Kalman filter (EKF) [30], which is an extension of Kalman–Bucy filter (1.16) to treat nonlinear dynamical systems with noise in the model and in the output

$$\begin{aligned} \dot{x} &= f(x) + w_x \\ y &= h(x) + w_y, \end{aligned} \tag{1.29}$$

where $w_x \in \mathbb{R}^n$ and $w_y \in \mathbb{R}^m$ represent Gaussian noise. It is assumed that system (1.27) is locally observable in the sense that the observability rank condition (1.26) is satisfied.

Note $\hat{x}$ as the estimation of x, then by linearizing the nonlinear functions f and h around its estimation, we have

$$\begin{aligned} A_t &= \frac{\partial f}{\partial x}|_{\hat{x}} \\ C_t &= \frac{\partial h}{\partial x}|_{\hat{x}}. \end{aligned}$$

Then the EKF is of the following form:

$$\begin{aligned} \dot{\hat{x}} &= f(\hat{x}) + K_t(y - h(\hat{x})) \\ \dot{P} &= A_t P + P A_t^T - K_t R_t K_t^T + Q_t \\ K_t &= P C_t^T R_t^{-1}, \end{aligned} \tag{1.30}$$

where Q_t and R_t are the driving noise covariance in the system dynamics and the measurement noise covariance, P is the estimation error covariance matrix and K_t is a dynamic gain. EKF is the most widely used estimation technique and has been validated by many different applications. However, since we linearize the nonlinearity of (1.29), the resulted EKF is therefore local, i.e., only local convergence can be assured, and it is impossible to have a global one. Interested readers are referred to [26] and references therein.

Nonlinear Luenberger observer and EKF are proposed for the general form (1.27), and the results are only local. To design global observers for nonlinear dynamical systems, we need to restrict our study to some special nonlinear forms (via some transformations).

Luenberger-Like Observer

The first special form of (1.27) is the following Lipschitz nonlinear dynamical system:

$$\begin{aligned} \dot{x} &= Ax + \varphi(x) \\ y &= Cx, \end{aligned} \tag{1.31}$$

where $\varphi(x)$ is a Lipschitz nonlinearity with a Lipschitz constant γ, i.e.,

$$||\varphi(x) - \varphi(\bar{x})|| \leq \gamma ||x - \bar{x}||. \tag{1.32}$$

Then we can design an observer of the following form:

$$\dot{\hat{x}} = A\hat{x} + \varphi(\hat{x}) + K[y - C\hat{x}], \tag{1.33}$$

where K represents the gain matrix and $\hat{x}$ means the estimated state. Thus, the estimation error dynamics is governed by

$$\dot{e} = (A - KC)e + [\varphi(x) - \varphi(\hat{x})], \tag{1.34}$$

where $e = x - \hat{x}$. The key point for such an observer is to determine K such that e tends to zero asymptotically.

In [44], the gain matrix K was chosen such that the following inequality is satisfied:

$$\gamma < \frac{\lambda_{min}(Q)}{2\lambda_{max}(P)}, \tag{1.35}$$

where K, Q and P are solutions of $(A - KC)^T P + P(A - KC) = -Q$, $\lambda_{min}(.)$ and $\lambda_{max}(.)$ represent the corresponding minimal and maximal eigenvalues. It has been proven in [44] that if this condition is fulfilled, then (1.33) is an asymptotic observer of (1.31).

Since the ratio in (1.35) is maximal when $Q = I$, the above problem is then reduced to choose K such that

$$\gamma < \frac{1}{2\lambda_{max}(P)}, \tag{1.36}$$

where $(A - KC)^T P + P(A - KC) = -I$.

Those results, however, are only useful to check the estimation convergence once the gain matrix K has been chosen, and no constructive method have been given to show how to choose such a K. To solve this problem, [39] has proposed to check, for some small ε, whether the following Riccati equation

$$AP + PA^T + P\left(\gamma^2 I - \frac{1}{\varepsilon}C^T C\right)P + I + \varepsilon I = 0 \tag{1.37}$$

has a positive definite solution P. If it does, then a choice of $K = \frac{1}{2\varepsilon}PC^T$ is shown to stabilize the error dynamics (1.34).

The condition (1.37) is only sufficient, and cannot be applied even for some cases where (A, C) is observable. Moreover, the stability of the observation error dynamics (1.34) intuitively depends on the eigenvalue of $(A - KC)$ and the Lipschitz constant γ, where the above condition does not clarify this relation. In order to understand this relation, [40] proved that (1.33) is an asymptotic observer of (1.27) if K is chosen such that $(A - KC)$ is Hurwitz and

$$\min_{\omega \in \mathbb{R}^+} \sigma_{\min}\left(A - KC - j\omega I\right) > \gamma,$$

where $\sigma_{\min}(.)$ represents the minimal singular values of the corresponding matrix.

Backstepping Observer

For system (1.27), if it is locally observable in the sense that the observability rank condition (1.26) is satisfied, then it is equivalent via a local diffeomorphism $\left[h, L_f h, \ldots, L_f^{n-1} h\right]^T$ to the following form:

$$\begin{aligned}
\dot{x}_1 &= \varphi(x) \\
\dot{x}_2 &= x_1 \\
&\vdots \\
\dot{x}_{n-1} &= x_{n-2} \\
\dot{x}_n &= x_{n-1} \\
y &= x_n,
\end{aligned} \tag{1.38}$$

where $\varphi(x)$ is smooth on $\mathscr{X}$, but not necessary to be Lipschitz.

In [26], a backstepping observer has been proposed which is of the following form:

$$\begin{aligned}
\dot{\hat{x}}_1 &= \varphi(\hat{x}) + K_1(\hat{x})\left(y - \hat{x}_n\right) \\
\dot{\hat{x}}_2 &= \hat{x}_1 + K_2(\hat{x})\left(y - \hat{x}_n\right) \\
&\vdots \\
\dot{\hat{x}}_{n-1} &= \hat{x}_{n-2} + K_{n-1}(\hat{x})\left(y - \hat{x}_n\right) \\
\dot{\hat{x}}_n &= \hat{x}_{n-1} + K_n(\hat{x})\left(y - \hat{x}_n\right) \\
y &= x_n,
\end{aligned} \tag{1.39}$$

where $K_i(\hat{x})$ are the gain functions to be determined. Note $e = [e_1, \ldots, e_n]^T$ with $e_i = x_i - \hat{x}_i$, then the observation error dynamics is governed by

$$
\begin{aligned}
\dot{e}_1 &= \varphi(x) - \varphi(\hat{x}) - K_1(\hat{x})e_n \\
\dot{e}_2 &= e_1 - K_2(\hat{x})e_n \\
&\vdots \\
\dot{e}_{n-1} &= e_{n-2} - K_{n-1}(\hat{x})e_n \\
\dot{e}_n &= e_{n-1} - K_n(\hat{x})e_n .
\end{aligned}
$$

A constructive approach based on backstepping technique has been proposed in [26] to calculate those gains $K_i(\hat{x})$ such that e might locally converge to zero in an asymptotic way. This method has been extended as well for multi-output systems in [26].

High-Gain Observer
Another special form of (1.31) is of the so-called triangular form

$$
\begin{aligned}
\dot{x}_1 &= \varphi_1(x_1, \ldots, x_n) \\
\dot{x}_2 &= x_1 + \varphi_2(x_2, \ldots, x_n) \\
&\vdots \\
\dot{x}_{n-1} &= x_{n-2} + \varphi_{n-1}(x_{n-1}, x_n) \\
\dot{x}_n &= x_{n-1} + \varphi_n(x_n) \\
y &= x_n ,
\end{aligned}
\tag{1.40}
$$

where f_i are smooth functions on $\mathscr{X}$ and globally Lipschitz. In [13], the following observer was proposed:

$$
\begin{aligned}
\dot{\hat{x}}_1 &= \varphi_1(\hat{x}_1, \ldots, \hat{x}_n) + \theta^n K_1 \left(y - \hat{x}_n\right) \\
\dot{\hat{x}}_2 &= \hat{x}_1 + \varphi_2(\hat{x}_2, \ldots, \hat{x}_n) + \theta^{n-1} K_2 \left(y - \hat{x}_n\right) \\
&\vdots \\
\dot{\hat{x}}_{n-1} &= \hat{x}_{n-2} + \varphi_{n-1}(\hat{x}_{n-1}, \hat{x}_n) + \theta^2 K_{n-1} \left(y - \hat{x}_n\right) \\
\dot{\hat{x}}_n &= \hat{x}_{n-1} + \varphi_n(\hat{x}_n) + \theta K_n \left(y - \hat{x}_n\right) ,
\end{aligned}
\tag{1.41}
$$

where

$$
K = [K_1, \ldots, K_n]^T ,
$$

with K_i being chosen such that $A - KC$ is Hurwitz. It has been proven that, for an enough large θ, the dynamics (1.41) is an exponential observer for the state of system (1.40). Since the gain is proportional of θ^n, large θ will yield high gain, and this is the reason why this kind of observer is named as high-gain observer [13]. Another important paper, published at the same time as [13], presented the same idea in [11], but with more attention to closed-loop control using high-gain observer where the famous peaking phenomenon has been investigated.

Adaptive Observer
Note that, for the special triangular form (1.40), high-gain observer utilizes a fixed constant gain, which in fact can be time varying. In [38], an adaptive law has been proposed for the triangular form (1.40) where the nonlinear function $\varphi_i(x_1, \ldots, x_i)$

could be non-Lipschitz, but was assumed to satisfy the following similar inequality

$$|\varphi_i(x_1,\ldots,x_i) - \varphi_i(x_1,\ldots,\bar{x}_i)| \leq \gamma(y)\left(|x_2 - \bar{x}_2| + \cdots + |x_i - \bar{x}_i|\right).$$

Based on this assumption, the following adaptive observer has been proposed:

$$\begin{aligned}
\dot{\hat{x}}_1 &= \varphi_1(\hat{x}_1,\ldots,\hat{x}_n) + \theta^n K_1\left(y - \hat{x}_n\right) \\
\dot{\hat{x}}_2 &= \hat{x}_1 + \varphi_2(\hat{x}_2,\ldots,\hat{x}_n) + \theta^{n-1} K_2\left(y - \hat{x}_n\right) \\
&\vdots \\
\dot{\hat{x}}_{n-1} &= \hat{x}_{n-2} + \varphi_{n-1}(\hat{x}_{n-1},\hat{x}_n) + \theta^2 K_{n-1}\left(y - \hat{x}_n\right) \\
\dot{\hat{x}}_n &= \hat{x}_{n-1} + \varphi_n(\hat{x}_n) + \theta K_n\left(y - \hat{x}_n\right) \\
\dot{\theta} &= l(\theta, y) \\
y &= x_1,
\end{aligned} \tag{1.42}$$

where θ is an extra state to update the gain, l is an $n+1$ times continuously differentiable function to be defined and K_i are the chosen constants such that there exist strictly positive real numbers q and a, and a symmetric matrix Q satisfying

$$\begin{aligned}
QP + P^T Q &\leq -aQ \\
qI \leq Q &\leq I,
\end{aligned}$$

where

$$P = \begin{pmatrix} 0 & 0 & \cdots & 0 & -K_1 \\ 1 & 0 & \cdots & 0 & -K_2 \\ \vdots & \vdots & \ddots & \vdots & \vdots \\ 0 & 0 & \cdots & 0 & -K_{n-1} \\ 0 & 0 & \cdots & 1 & -K_n \end{pmatrix}. \tag{1.43}$$

By properly choosing the function l, it has been proven in [38] that the state of the dynamics (1.42) can asymptotically converge to that of (1.40) even if its nonlinearity does not satisfy the Lipschitz condition.

Observer Normal Forms

We have reviewed some types of nonlinear observers for the general nonlinear form (1.27) and also for some special forms, and those techniques synthesize observers directly for the studied nonlinear dynamical systems. Besides, there exists as well an indirect way to design nonlinear observers, which is based on the transformation technique. The basic idea of such a technique is to transform the original system into a simple observer normal form via a change of coordinates (a diffeomorphism). The advantage of this method is that, by well choosing the desired simple observer normal form, we can reuse the existing observers proposed in the literature to estimate the state of the transformed observer normal form, and then obtain the state estimation for the original system by inverting the deduced diffeomorphism. The literature about this technique is vast. Since the pioneer works of [3, 25] for single-output dynamical systems and [27, 47] for the case of multi-input multi-output dynamical systems, many other results have been published by following the same idea.

The goal of this monograph is to present how this method can be used to design observers for different nonlinear dynamical systems. Compared to the existing books on observer design, the peculiarity of this book is the idea of keeping things simple, and focuses only on one approach: differential geometry. It can be regarded as a complementary of those existing books on this topic, since no books have investigated the observer design problem via differential geometric approach.

To apply this technique, two important issues need to be taken into account:

- Choice of the targeted normal form for which an observer can be easily designed via existing results;
- Deduction of the diffeomorphism which can transform the studied system into the target normal form.

Obviously, given a nonlinear dynamical system (1.27), different choices of targeted normal forms require to deduce different diffeomorphism. Intuitively, the larger class of targeted normal forms we choose, the more difficult to deduce the corresponding diffeomorphism.

Observer Normal Form with Output Injection

Let us start with the simplest normal form. As we have seen for linear time-invariant dynamical system (1.15), we can always find a change of coordinates such that (1.15) can be transformed into the observable canonical form (1.18). It is natural to ask the question whether we can extend the same solution to nonlinear dynamical systems. In other words, the first targeted normal form for nonlinear dynamical systems, similar to (1.18) for linear case, is of the following form:

$$\begin{aligned} \dot{z} &= A_O z + \beta(y) \\ y &= C_O z, \end{aligned} \tag{1.44}$$

where

$$A_O = \begin{pmatrix} 0 & \cdots & 0 & 0 & 0 \\ 1 & \cdots & 0 & 0 & 0 \\ \vdots & \ddots & \ddots & \cdots & \vdots \\ 0 & \cdots & 1 & 0 & 0 \\ 0 & \cdots & 0 & 1 & 0 \end{pmatrix}, \quad C_O = \begin{pmatrix} 0 & \cdots & 0 & 1 \end{pmatrix} \tag{1.45}$$

and $\beta(y)$ is only a function of the measurable output y. It is clear that, compared to (1.18), the main difference for nonlinear case is that the term $\beta(y)$ admits a nonlinear function of y, while it is only linear in (1.18). This normal form, firstly proposed in [25], is named as observer normal form with output injection.

Observer Normal Form with Output Injection and Output Diffeomorphism

We can notice that the output y stays as linear function of the transformed state in the first normal form (1.44). To extend this normal form to a more general one, the second normal form enables the nonlinear output, which can be written as follows:

$$\begin{aligned} \dot{z} &= A_O z + \beta(y) \\ y &= \varphi(C_O z), \end{aligned} \tag{1.46}$$

where A_O and C_O are defined in (1.45), and $\varphi(C_O z)$ might be nonlinear. Note that if $\varphi(C_O z) = C_O z$, then the normal form (1.46) is equivalent to (1.44).

Observer Normal Form with Output Injection by Means of Extended Dynamics
Note that the first and the second targeted normal forms have the same dimension as that of (1.27), and therefore the deduced transformation is of dimension n as well. A larger class of normal forms is naturally to consider the higher dimensional normal forms.

Denote $n + m$ as the dimension of the new targeted normal form, then the transformation is not anymore a diffeomorphism, but a differential injective mapping from $\mathbb{R}^n$ to $\mathbb{R}^{n+m}$. One solution to keep the deduced transformation as a diffeomorphism is the so-called immersion technique. By adding an auxiliary dynamics of dimension m into (1.27), this leads to the following augmented system:

$$\begin{aligned} \dot{x} &= f(x) \\ \dot{w} &= \eta(w, y) \\ y &= h(x), \end{aligned} \tag{1.47}$$

where $w \in \mathbb{R}^m$ is the artificial dynamics which needs to be determined. We would like to emphasize that w can be seen as known variables, since its dynamics is only determined by y and we can freely choose the initial condition of w. Using immersion, the larger targeted normal form can be given by the following form:

$$\begin{aligned} \dot{z} &= A_O z + \beta(y, w) \\ \dot{\xi} &= \mu(\xi, y) \\ y &= \varphi(C_O z), \end{aligned} \tag{1.48}$$

where $\xi \in \mathbb{R}^m$ is the corresponding transformed variables in the normal form, and $\varphi(.)$ is the function of the output which needs to be determined, similar as the output of the second normal form (1.46).

Output-Depending Observer Normal Form
In the first three targeted normal forms, the linear part is always of the Brunovsky form A_O. Such a choice is for the purpose of guaranteeing the observability. Note that the output y is also measurable, from observability point of view, and this linear term A_O might depend as well on y. Motivated by this, a larger class of targeted normal forms can be chosen as follows:

$$\begin{aligned} \dot{z} &= A_O(y) z + \beta(y) \\ y &= C_O z, \end{aligned} \tag{1.49}$$

where the matrix $A_O(y)$ is of the following form:

$$A_O(y) = \begin{pmatrix} 0 & \cdots & 0 & 0 & 0 \\ \alpha_1(y) & \cdots & 0 & 0 & 0 \\ \vdots & \ddots & \vdots & \cdots & \vdots \\ 0 & \cdots & \alpha_{n-2}(y) & 0 & 0 \\ 0 & \cdots & 0 & \alpha_{n-1}(y) & 0 \end{pmatrix}, \text{ and } C_O = \begin{pmatrix} 0 \cdots 0 \; 1 \end{pmatrix}, \tag{1.50}$$

with $\alpha_i(y) \neq 0, \forall y \in \mathbb{R}^m$ for $1 \leq i \leq n-1$ are functions of y to be determined. Due to the fact that $A_O(y)$ now is dependent on the output, we name this normal form (1.49) as output-depending observer normal form. It is obvious that if $\alpha_i(y) = 1$, then this normal form is equivalent to (1.44).

Output-Depending Observer Normal Form and Output Diffeomorphism
Following the same idea to extend the first targeted normal form (1.44) to (1.46) by applying an output transformation, we can make the same generalization for the output-depending observer normal form (1.49), which permits us to propose the following normal form:

$$\begin{aligned} \dot{z} &= A_O(y)z + \beta(y) \\ y &= \varphi(C_O z), \end{aligned} \tag{1.51}$$

where $A_O(y)$ and C_O are defined in (1.50).

Output-Depending Observer Normal Form by Means of Extended Dynamics
To have a larger class of normal form than (1.51), we can also apply the immersion technique to generalize the output-depending observer normal form, which is of the following form:

$$\begin{aligned} \dot{z} &= A_O(y)z + \beta(y, w) \\ \dot{\xi} &= \mu(\xi, y) \\ y &= \varphi(C_O z), \end{aligned} \tag{1.52}$$

where $A_O(y)$ and C_O are defined in (1.50).

Up to now, we have presented six types of observer normal forms. After having chosen the targeted normal form for nonlinear dynamical system (1.27), the key work is to seek for sufficient (and necessary if they exist) conditions such that there exists a local diffeomorphism which can transform system (1.27) into the chosen targeted normal form.

To deduce the desired diffeomorphism, theoretically there are two different methods based either on differential algebra or on differential geometry. As indicated in the title of this book, we are interested in solving the observer design problem for nonlinear dynamical systems via transformation, and will focus only on the differential geometric method to deduce the desired diffeomorphism. Precisely, the next chapter will present the basic background of differential geometry which will be used in Chaps. 3–9. Then, from Chaps. 3–9, we would like to present how to deduce the diffeomorphism and how to design observers for different targeted normal forms, respectively.

References

1. Belevitch, V.: Classical Network Theory, vol. 7. Holden-Day, San Francisco (1968)
2. Besançon, G.: Nonlinear Observers and Applications, vol. 363. Springer, Berlin (2007)
3. Bestle, D., Zeitz, M.: Canonical form observer design for non-linear time-variable systems. Int. J. Control **38**(2), 419–431 (1983)
4. Bhat, S., Bernstein, D.: Finite-time stability of continuous autonomous systems. SIAM J. Control Optim. **38**(3), 751–766 (2000)
5. Bhattacharyya, S.: Observer design for linear systems with unknown inputs. IEEE Trans. Autom. Control **23**(3), 483–484 (1978)
6. Chen, Y., Vinagre, B.M.: A new IIR-type digital fractional order differentiator. Signal Process. **83**(11), 2359–2365 (2003)
7. Conte, G., Moog, C., Perdon, A.M.: Algebraic Methods for Nonlinear Control Systems. Springer Science & Business Media, New York (2007)
8. Dabroom, A., Khalil, H.: Numerical differentiation using high-gain observers. In: Proceedings of the 36th IEEE Conference on Decision and Control, vol. 5, pp. 4790–4795 (1997)
9. Diop, S., Fliess, M.: Nonlinear observability, identifiability, and persistent trajectories. In: Proceedings of the 30th IEEE Conference on Decision and Control, pp. 714–719 (1991)
10. Engel, R., Kreisselmeier, G.: A continuous-time observer which converges in finite time. IEEE Trans. Autom. Control **47**(7), 1202–1204 (2002)
11. Esfandiari, F., Khalil, H.: Output feedback stabilization of fully linearizable systems. Int. J. Control **56**(5), 1007–1037 (1992)
12. Fliess, M.: Some remarks on nonlinear invertibility and dynamic state feedback. Theory Appl. Nonlinear Control Syst. **8**, 115–121 (1986)
13. Gauthier, J.P., Hammouri, H., Othman, S.: A simple observer for nonlinear systems applications to bioreactors. IEEE Trans. Autom. Control **37**(6), 875–880 (1992)
14. Guo, B.Z., Zhao, Z.L.: On convergence of tracking differentiator. Int. J. Control **84**(4), 693–701 (2011)
15. Hautus, M.: Strong detectability and observers. Linear Algebra Appl. **50**, 353–368 (1983)
16. Hermann, R., Krener, A.: Nonlinear controllability and observability. IEEE Trans. Autom. Control **22**(5), 728–740 (1977)
17. Hespanha, J.: Linear Systems Theory. Princeton University Press, Princeton (2018)
18. Ibrir, S.: Online exact differentiation and notion of asymptotic algebraic observers. IEEE Trans. Autom. Control **48**(11), 2055–2060 (2003)
19. Isidori, A.: Nonlinear Control Systems. Communications and Control Engineering, 3rd edn. Springer, London (1995)
20. Kalman, R.: A new approach to linear filtering and prediction problems. J. Basic Eng. **82**(1), 35–45 (1960)
21. Kalman, R.E., Bucy, R.S.: New results in linear filtering and prediction theory. J. Basic Eng. **83**(1), 95–108 (1961)
22. Kazantzis, N., Kravaris, C.: Nonlinear observer design using Lyapunov's auxiliary theorem. Syst. Control Lett. **34**(5), 241–247 (1998)
23. Kreisselmeier, G.: Adaptive observers with exponential rate of convergence. IEEE Trans. Autom. Control **22**(1), 2–8 (1977)
24. Kreisselmeier, G., Engel, R.: Nonlinear observers for autonomous Lipschitz continuous systems. IEEE Trans. Autom. Control **48**(3), 451–464 (2003)
25. Krener, A., Isidori, A.: Linearization by output injection and nonlinear observers. Syst. Control Lett. **3**(1), 47–52 (1983)
26. Krener, A., Kang, W.: Locally convergent nonlinear observers. SIAM J. Control Optim. **42**(1), 155–177 (2003)
27. Krener, A., Respondek, W.: Nonlinear observers with linearizable error dynamics. SIAM J. Control Optim. **23**(2), 197–216 (1985)
28. Krener, A., Xiao, M.: Nonlinear observer design in the Siegel domain. SIAM J. Control Optim. **41**(3), 932–953 (2002)

29. Levant, A.: Robust exact differentiation via sliding mode technique. Automatica **34**(3), 379–384 (1998)
30. Ljung, L.: Asymptotic behavior of the extended Kalman filter as a parameter estimator for linear systems. IEEE Trans. Autom. Control **24**(1), 36–50 (1979)
31. Ljung, L.: System Identification: Theory for the User. Prentice Hall, Hoboken (1987)
32. Luenberger, D.: An introduction to observers. IEEE Trans. Autom. Control **16**(6), 596–602 (1971)
33. Lynch, A., Bortoff, S.: Nonlinear observers with approximately linear error dynamics: the multivariable case. IEEE Trans. Autom. Control **46**(6), 927–932 (2001)
34. Menard, T., Moulay, E., Perruquetti, W.: A global high-gain finite-time observer. IEEE Trans. Autom. Control **55**(6), 1500–1506 (2010)
35. Molinari, B.: A strong controllability and observability in linear multivariable control. IEEE Trans. Autom. Control **21**(5), 761–764 (1976)
36. Perruquetti, W., Floquet, T., Moulay, E.: Finite-time observers: application to secure communication. IEEE Trans. Autom. Control **53**(1), 356–360 (2008)
37. Phelps, A.: On constructing nonlinear observers. SIAM J. Control Optim. **29**(3), 516–534 (1991)
38. Praly, L.: Asymptotic stabilization via output feedback for lower triangular systems with output dependent incremental rate. IEEE Trans. Autom. Control **48**(6), 1103–1108 (2003)
39. Raghavan, S., Hedrick, K.: Observer design for a class of nonlinear systems. Int. J. Control **59**(2), 515–528 (1994)
40. Rajamani, R.: Observers for Lipschitz nonlinear systems. IEEE Trans. Autom. Control **43**(3), 397–401 (1998)
41. Rugh, W.: Linear System Theory, vol. 2. Prentice Hall, Upper Saddle River (1996)
42. Shen, Y., Xia, X.: Semi-global finite-time observers for nonlinear systems. Automatica **44**(12), 3152–3156 (2008)
43. Shen, Y., Huang, Y., Gu, J.J.: Global finite-time observers for Lipschitz nonlinear systems. IEEE Trans. Autom. Control **56**(2), 418–424 (2011)
44. Thau, F.: Observing the state of non-linear dynamic systems. Int. J. Control **17**(3), 471–479 (1973)
45. Trentelman, H., Stoorvogel, A., Hautus, M.: Control Theory for Linear Systems. Springer Science & Business Media, New York (2012)
46. Utkin, V.: Sliding Modes in Control and Optimization. Springer Science & Business Media, New York (2013)
47. Xia, X.H., Gao, W.B.: Nonlinear observer design by observer error linearization. SIAM J. Control Optim. **27**(1), 199–216 (1989)

Chapter 2
Background on Differential Geometry

Abstract Through this book, we will adopt a geometric point of view to analyze nonlinear dynamical systems. Geometrically speaking, the behavior of a dynamical system can be governed by the associated vector field. In this sense, the dynamical behavior of such a system can be presented by smooth trajectories tangent to this vector field. Then, for a given dynamical system endowed with measurements, we can use the differential forms to analyze its property of observability. In order to make this book be self-contained, this chapter is devoted to presenting some necessary elements (vector fields, differential forms, Lie bracket, Poincaré's Lemma) of differential geometry which will be used when investigating nonlinear observer normal forms.

2.1 Vector Fields: Derivation and Dynamics

Denote $\mathscr{X}$ as a neighborhood of 0 in $\mathbb{R}^n$ with coordinates $x = (x_1, \ldots, x_n)$ in $\mathscr{X}$, and note $T_x\mathscr{X}$ as the *tangent space* to $\mathscr{X}$ at x. If we denote by $v_x \in T_x\mathscr{X}$ a vector tangent to $\mathscr{X}$ at x, then the following space [4, 8]

$$T\mathscr{X} = \{v_x : x \in \mathscr{X}\} \tag{2.1}$$

is called the *tangent fiber bundle* of $\mathscr{X}$, which endows with the following projection mapping

$$p : T\mathscr{X} \to \mathscr{X},$$

which assigns a tangent vector v_x to its base point x.

The *fiber* of $x \in \mathscr{X}$ is just the inverse image (or preimage) of x: $p^{-1}(x) = T_x\mathscr{X}$, which is the tangent space to $\mathscr{X}$ at x. A *vector field* f is an assignment of a tangent vector for each point $x \in \mathscr{X}$. Thus, a vector field f is a section of this bundle, i.e.,

$$f : \mathscr{X} \to T\mathscr{X}$$

D. Boutat and G. Zheng, *Observer Design for Nonlinear Dynamical Systems*,
Lecture Notes in Control and Information Sciences 487,
https://doi.org/10.1007/978-3-030-73742-9_2

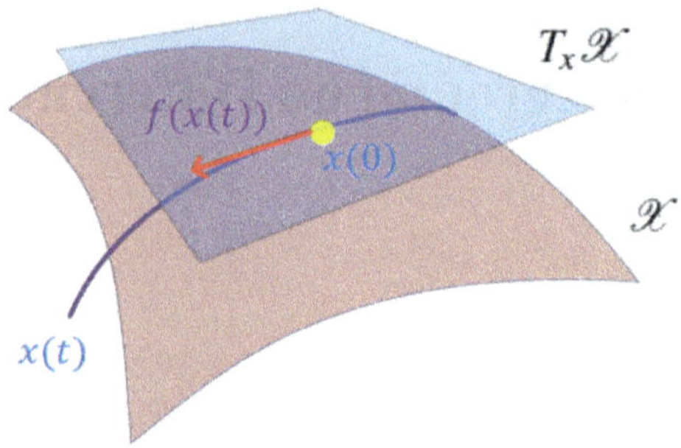

Fig. 2.1 Tangent space

such that $p \circ f = Id_{\mathscr{X}}$ is the identity of $\mathscr{X}$. By abuse of language, the space $T\mathscr{X}$ is often called the tangent bundle of $\mathscr{X}$ without mentioning the projection mapping p (Fig. 2.1).

For the coordinates $x \in \mathscr{X}$, a vector field $f(x)$ can be represented as follows:

$$f(x) = (b_1(x), \ldots, b_n(x))^T = \sum_{i=1}^{n} b_i(x) \frac{\partial}{\partial x_i}, \tag{2.2}$$

where $b_i(x)$ for $1 \leq i \leq n$ is a function on $\mathscr{X}$ and $\frac{\partial}{\partial x_i}$ denotes the partial derivative in the direction of $e_i = \left(0, \ldots, 0, \underbrace{1}_{i\text{th component}}, 0, \ldots, 0\right)^T$.

The vector field $f(x)$ defined in (2.2) has two interpretations. Firstly, it can be seen as a directional derivation. Precisely, consider a smooth function $h(x)$ on $\mathscr{X}$, then its differential $dh = \left(\frac{\partial h}{\partial x_1}, \ldots, \frac{\partial h}{\partial x_n}\right)$ can be evaluated on $f(x)$ as follows:

$$dh(f) = \sum_{i=1}^{n} b_i(x) \frac{\partial h}{\partial x_i}. \tag{2.3}$$

This expression is known as the derivative of the function $h(x)$ in the direction of $f(x)$. It is called the *Lie derivative* of $h(x)$ along $f(x)$ and is noted as

$$L_f h = \sum_{i=1}^{n} b_i(x) \frac{\partial h}{\partial x_i}. \tag{2.4}$$

Example 2.1 Let $f(x) = \frac{\partial}{\partial x_i}$, then the Lie derivative of a function $h(x)$ in the direction of f is the well-known partial derivative of h in the direction of x_i. Thus, we have

$$L_f h = \frac{\partial h}{\partial x_i}.$$

□

Example 2.2 Consider the vector field $f(x) = x_2\frac{\partial}{\partial x_1} - x_1\frac{\partial}{\partial x_2}$,

- for $h(x) = x_1$, we have $L_f h = x_2$;
- for $h(x) = x_1^2 + x_2^2$, we have $L_f h = 0$. This implies that the derivative of $h(x)$ in the direction of $f(x)$ is zero, i.e., the function h is constant in the direction of f.

□

Example 2.3 Given two vector fields $f_1(x)$ and $f_2(x)$, and two functions $b_1(x)$ and $b_2(x)$ for $x \in \mathscr{X}$. Note $f(x) = b_1(x)f_1(x) + b_2(x)f_2(x)$, then we can show that

$$L_f h = b_1 L_{f_1} h + b_2 L_{f_2} h.$$

The detailed deduction is left to the reader. □

The second interpretation of a vector field f is that it governs a dynamical system as follows:

$$\dot{x}(t) = f(x(t)), \tag{2.5}$$

where $\dot{x}(t) = \frac{dx(t)}{dt}$ is the derivative of the solution curve $x(t)$ with respect to time t, and the vector field f is tangent to curve $x(t)$. The phase portrait of dynamical system (2.5) is the family of solution curves that fill the entire space where f is defined.

For the constant case where $f = \frac{\partial}{\partial x_i}$, the phase portrait of $\dot{x}(t) = f(x(t))$ consists of the parallel lines to the direction $\frac{\partial}{\partial x_i}$. Thus, the line $x_j = c_j$ for $j \neq i$ where c_j are constants. In the plane $\mathbb{R}^2$ with coordinates (x_1, x_2), the phase portrait for $f = \frac{\partial}{\partial x_1}$ consists of the straight horizontal lines, and for $f = \frac{\partial}{\partial x_2}$ it consists of the straight vertical lines (see Fig. 2.2).

Let us consider another example in the plane $\mathbb{R}^2$, which has the following vector field

$$f = x_2\frac{\partial}{\partial x_1} - x_1\frac{\partial}{\partial x_2},$$

then the phase portrait of this dynamical system governed by f is origin-centered circles. Of course, the origin is also a solution where the circle curve reduces to a point (see Fig. 2.3).

Another concrete example is the harmonic motion of a pendulum given by Newton's second law

$$m\frac{d^2\theta}{dt^2} + \frac{g}{L}\sin(\theta) = 0, \tag{2.6}$$

where m is the mass at the end of the pendulum, g is the gravity constant, L is the length of string and θ is the angle between the string position at a given time to its position at rest. To rewrite (2.6) into the state-space representation, we set $x_1 = \theta$ and $x_2 = \dot{\theta}$. Then we have the following dynamical system:

$$\begin{aligned} \dot{x}_1 &= x_2 \\ \dot{x}_2 &= -\tfrac{g}{Lm}\sin(x_1), \end{aligned} \tag{2.7}$$

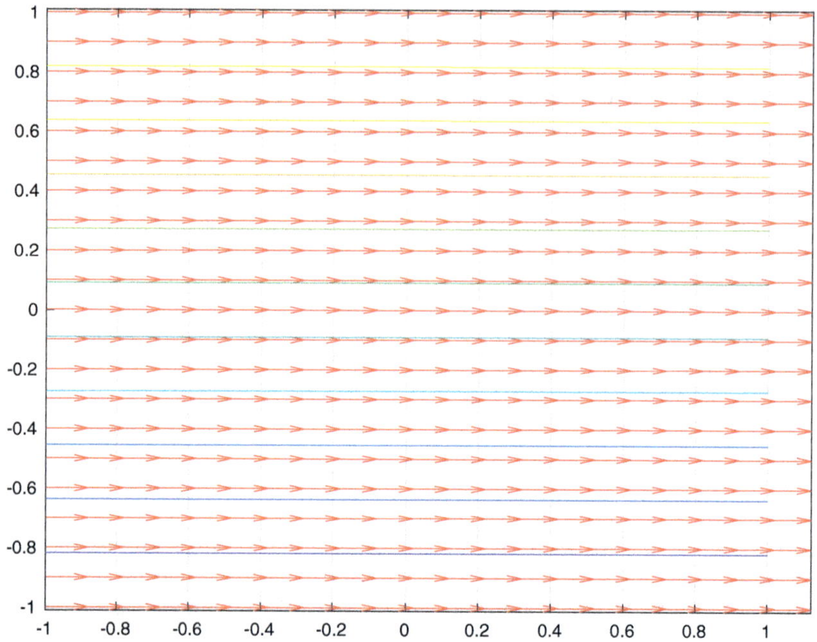

Fig. 2.2 Phase portrait of constant vector field $\frac{\partial}{\partial x_1}$

which can be regarded as a dynamics generated by the following vector field:

$$f = x_2 \frac{\partial}{\partial x_1} - \frac{g}{Lm} s \sin(x_1) \frac{\partial}{\partial x_2}.$$

One way to draw the phase portrait of this dynamical system is the notion of first integral or constant of motion in mechanics. First integral is the total energy conservation from Newton's second law. To achieve this, we multiply (2.6) by $\dot{\theta}$ to get the following equation:

$$\frac{d}{dt}\left(m\frac{\theta^2}{2} + 1 - \frac{g}{L}\cos(\theta) \right) = 0. \tag{2.8}$$

Integrating the above equation shows that the energy $E = E_c + E_p = m\frac{\theta^2}{2} + 1 - \frac{g}{L}\cos(\theta)$ is conserved, where $E_c = m\frac{\theta^2}{2}$ represents the kinetic energy and $E_p = 1 - \frac{g}{L}\cos(\theta)$ is the potential energy.

Now let $f = x_2 \frac{\partial}{\partial x_1} - \frac{g}{Lm}\sin(x_1)\frac{\partial}{\partial x_2}$, then $L_f E = 0$ that means E is constant along the direction of f. Therefore, the curve solutions of (2.7) are just the level curves of the energy E, which are curves at a fixed height $E = c$ constant as shown in Fig. 2.3.

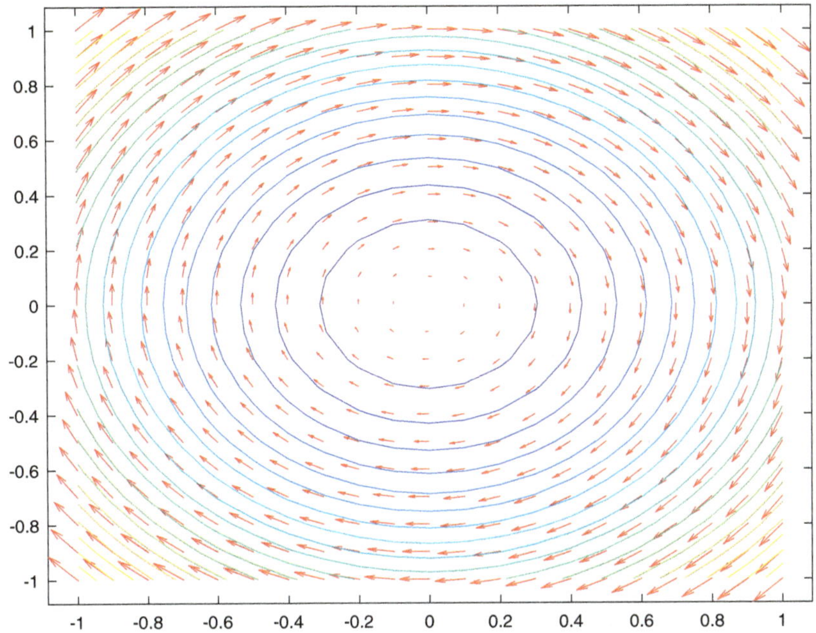

Fig. 2.3 Phase portrait of f

2.2 Lie Bracket of Vector Fields

Given a smooth manifold, a vector field defines the direction in which the trajectories of a dynamical system should evolve. It is known that we can assign a new vector field by just combining those existing vector fields, which is named as Lie bracket of vector fields [7].

To be more precise, let us consider $f = \sum_{i=1}^{n} f_i \frac{\partial}{\partial x_i}$ and $g = \sum_{i=1}^{n} g_i \frac{\partial}{\partial x_i}$, two vector fields on an open set of $\mathbb{R}^n$ where f_i and g_i are smooth functions, then the Lie bracket of these two vector fields is defined below.

Definition 2.1 The Lie bracket of f and g is a vector field defined as

$$[f, g] = \sum_{i=1}^{n} (L_f g_i - L_g f_i) \frac{\partial}{\partial x_i}, \tag{2.9}$$

where $L_f g_i$ and $L_g f_i$ are the Lie derivatives of functions g_i and f_i, respectively, in the directions of f and g.

The introduction of Lie bracket of vector fields is of particular interest in many domains, such as in foliation theory where it was used to show the integrability, and in control theory where it was used to study the controllability and observability. Another motivation to introduce the Lie bracket of vector fields is to judge whether two vector fields can be locally transformed, by means of a change of coordinates, into two constant vector fields.

Example 2.4 Consider in the space $\mathbb{R}^3$ the following two vector fields:

$$f_1 = -2x_2\frac{\partial}{\partial x_1} + \frac{\partial}{\partial x_2}$$

and

$$f_2 = \left(2x_2e^{x_3} - 1\right)\frac{\partial}{\partial x_1} - e^{x_3}\frac{\partial}{\partial x_2} + \frac{\partial}{\partial x_3}.$$

By applying the following change of coordinates

$$\phi(x_1, x_2, x_3) = \begin{pmatrix} x_1 + x_2^2 + x_3 \\ x_2 + e^{x_3} \\ x_3 \end{pmatrix} = \begin{pmatrix} z_1 \\ z_2 \\ z_3 \end{pmatrix}$$

a straightforward calculation leads to the Jacobian of ϕ as follows:

$$\phi_* = \begin{pmatrix} 1 & 2x_2 & 1 \\ 0 & 1 & e^{x_3} \\ 0 & 0 & 1 \end{pmatrix},$$

which transforms the two vector fields into $\phi_*(f_1) = \frac{\partial}{\partial z_2}$ and $\phi_*(f_2) = \frac{\partial}{\partial z_3}$ in the z_i coordinates. Such a transformation exists because of $[f_1, f_2] = 0$. We will explain this in detail by using the notion of integrability on 1-forms. □

The Lie bracket can also be used to measure the non-commutativity of flows along these vector fields. Starting from a point x_0, let us track the infinitesimal flow ϕ_t^f (trajectory) of f during a small time interval t, then ϕ_t^g of g for the same time interval, then that of $-f$ and finally that of $-g$. It can be shown that, after $4t$, the trajectory is

Fig. 2.4 Infinitesimal description of Lie bracket

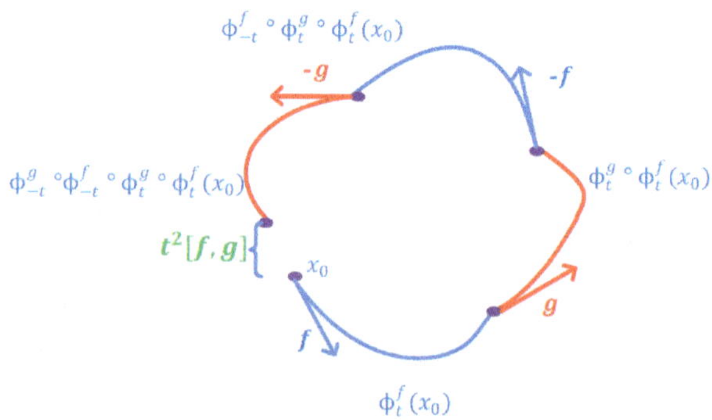

$$x(4t) = \phi^g_{-t} \circ \phi^f_{-t} \circ \phi^g_t \circ \phi^f_t(x_0) = x(0) + t^2[f, g] + \mathscr{R}_{es}, \tag{2.10}$$

where $\mathscr{R}_{es}$ contains Lie bracket involving in the direction of $[f, g]$. Therefore, we have $x(4t) = x(0)$ if and only if $[f, g] = 0$, which is equivalent to the condition that the vector fields f and g commute. Obviously, concerning the flow of dynamical system, $[f, g]$ can be seen as the obstruction to go back to the starting point (see Fig. 2.4).

Example 2.5 Consider the two linear vector fields $f = Ax$ and $g = Bx$, it is easy to calculate that

$$[f, g] = (BA - AB)x.$$

Thus $[f, g] = 0$ if and only if $BA = AB$, which implies that matrices A and B commute. The commutativity condition $BA = AB$ can also be obtained by checking (2.10)

$$x(4t) = e^{-tA}e^{-tB}e^{tA}e^{tB}x(0),$$

which is equal to $x(0)$ for all t if and only if $AB = BA$. □

Since the Lie bracket of vector fields enables us to assign a new vector field, the flow of dynamical system might evolve in the direction of iterative Lie bracket of vector fields. In order to simplify the notations, the symbol *ad* (adjoint endomorphism) was introduced in the literature to express the successive Lie bracket. Given two vector fields f_1 and f_2, their Lie bracket is noted as

$$ad_{f_1} f_2 := [f_1, f_2].$$

Iteratively, for all $i \geq 0$, the successive Lie bracket can be written as

$$ad^{i+1}_{f_1} f_2 = \left[f_1, ad^i_{f_1} f_2\right],$$

where $ad^0_{f_1} f_2 = f_2$ by convention.

It is worth highlighting that the Lie bracket is an intrinsic property in the sense that it is independent of the choice of change of coordinates. More precisely, given two vector fields f_1 and f_2 of dimension n, if $z = \phi(x) = (\phi_1(x), \ldots, \phi_n(x))^T$ is a change of coordinates (diffeomorphism), then we have

$$\phi_*([f_1, f_2]) = [\phi_*(f_1), \phi_*(f_2)], \tag{2.11}$$

where $\phi_*(f_k) = \frac{\partial \phi}{\partial x}(f_k)$ for $k = 1, 2$ represents the transformation of vector field f_k by means of the Jacobian of ϕ defined as follows:

$$\phi_* := \frac{\partial \phi}{\partial x} = \left(\frac{\partial \phi_i}{\partial x_j}\right)_{1 \leq i, j \leq n}.$$

2.3 Differential Forms

Based on the definition of tangent fiber bundle $T\mathscr{X}$ given in (2.1), the dual of the tangent fiber bundle $T\mathscr{X}$ is the cotangent fiber bundle [2]

$$p^* : T^*\mathscr{X} \to \mathscr{X},$$

whose elements $\theta_x \in T^*\mathscr{X}$ at a point x are linear forms

$$\theta_x : T_x\mathscr{X} \to \mathbb{R}.$$

Then, any section θ of the cotangent bundle is called a differential 1-form.

From the differential calculus point of view, the product of the differential dx_i of the ith coordinate function x_i with $\frac{\partial}{\partial x_j}$ is defined as follows:

$$dx_i\left(\frac{\partial}{\partial x_j}\right) = \frac{\partial x_i}{\partial x_j} = \delta^i_j,$$

where δ^i_j denotes the Kronecker symbol, i.e., $\delta^i_j = 1$ when $i = j$, otherwise it is equal to 0 for $i \neq j$.

In the local coordinate, a differential 1-form can be generally written as

$$\theta = \sum_{i=1}^{n} a_i(x)dx_i, \tag{2.12}$$

where $a_i(x)$ is a function of $x \in \mathscr{X}$ for $1 \leq i \leq n$. The evaluation of the differential 1-form θ given in (2.12) on the vector field f given in (2.2) is a function $\theta(f)$ on $\mathscr{X}$ defined by

$$\theta(f)(x) = \sum_{i=1}^{n} a_i(x)b_i(x),$$

which can also be regarded as the interior product of f by θ.

As a simple example of interior product, consider $\theta = dh(x)$ which is the differential of function $h : \mathscr{X} \to \mathbb{R}$, noted as

$$dh = \sum_{i=1}^{n} \frac{\partial h}{\partial x_i} dx_i,$$

then the interior product of f by θ can be written as

$$i_f\theta = dh(f) = \sum_{i=1}^{n} \frac{\partial h_i}{\partial x_i} f_i.$$

It can be seen clearly that the interior product of f by dh is exactly the Lie derivative of h along the vector field f, i.e.,

$$L_f h = i_f dh.$$

After having introduced the notation of Lie bracket and differential 1-form, it is interesting to investigate the relationship between them [1, 10].

Let us recall some well-known basic facts from differential geometry. The first fact is the so-called Cartan's identity. It gives the relation between the Lie derivative L_X in the direction of a vector field X, the differential d and the interior product $\imath_X$ as follows:

$$L_X = \imath_X d + d\imath_X, \tag{2.13}$$

where, for the 1-form ν, we have $\imath_X \nu = \nu(X)$ which is the evaluation of ν on X with $\imath_X q = 0$ for a function q (called 0-form or form of degree 0).

The second fact is the wedge product of two 1-forms ν_1 and ν_2 which is denoted by $\nu_1 \wedge \nu_2$. It is a differential 2-form defined by its evaluation on two vector fields X and Y as follows:

$$\nu_1 \wedge \nu_2(X, Y) = \nu_1(X)\nu_2(Y) - \nu_1(Y)\nu_2(X). \tag{2.14}$$

The final point known as the Jacobi's identity for three vector fields X, Y and Z is given by

$$[[X, Y], Z] + [[Z, X], Y] + [[Y, Z], X] = 0. \tag{2.15}$$

A 2-form ω is a linear combination of element $\omega_1 \wedge \omega_2$ where ω_i are 1-forms. To know how to evaluate ω on two vector fields f_1 and f_2, it is sufficient to know that $\omega_1 \wedge \omega_2(f_1, f_2) = \omega_1(f_1)\omega_2(f_2) - \omega_1(f_2)\omega_2(f_1)$. This rule of calculation is linked to the Lie derivative and the Lie bracket, and it is stated in the following proposition.

Proposition 2.1 *The differential of a differential* 1*-form* ω *is a differential* 2*-form whose evaluation on two vector fields* f_1 *and* f_2 *is computed as follows:*

$$d\omega(f_1, f_2) = L_{f_1}\omega(f_2) - L_{f_2}\omega(f_1) - \omega([f_1, f_2]), \tag{2.16}$$

where $L_{f_i}\omega(f_j)$ *is the Lie derivation of the function* $\omega(f_j)$ *in the direction of vector field* f_i *for* $1 \leq i, j \leq 2$.

Definition 2.2 A differential 1-form ω is said to be closed if $d\omega = 0$.

Therefore, if a differential 1-form ω is closed, then it can be expressed locally as

$$d\omega\left(\frac{\partial}{\partial x_i}, \frac{\partial}{\partial x_j}\right) = 0$$

for all $1 \leq i, j \leq n$. In other words, the evaluation of ω on the basis of vector fields is 0.

A typical example of closed 1-form is the differential of a function $\omega = dh$, for which it is easy to see that $d\omega = 0$. The question is raised whether $d\omega = 0$ implies $\omega = dh$? And this result is provided by the following Poincaré's lemma.

Lemma 2.1 (Poincaré's Lemma) *Given a differential* 1*-form* ω *on* $\mathscr{X}$ *with* $d\omega = 0$, *then* ω *is exact on a contractile open subset* $\mathscr{V}$ *of* $\mathscr{X}$. *Moreover, there exists a function* h *such that* $\omega = dh$ *on* $\mathscr{V}$, *where* h *is defined as follows:*

$$h(x) = \int_\gamma \omega = \int_0^1 \omega(\gamma'(t))dt,$$

where γ *is any smooth path* $\gamma : [0, 1] \to \mathscr{V}$ *with* $\gamma(0) = 0$ *and* $\gamma(1) = x$.

It is worth noting that the function h is well defined because the integral does not depend on the choice of the path γ that links $\gamma(0)$ to $\gamma(1)$. Indeed, if γ_1 is another path, then we have $\int_\gamma \theta - \int_{\gamma_1} \theta = \int_{\gamma_2} \theta$ where γ_2 is the closed loop obtained by following firstly γ and then $-\gamma_1$. As this closed loop borders a contractile domain $\mathscr{D} \subset \mathscr{V}$, then thanks to Stokes' theorem, we obtain $\int_{\gamma_2} \theta = \int_{\mathscr{D}} d\theta = 0$.

We should pay attention as well to the hypotheses of the above lemma, i.e., the domain must be contractile and the 1-form must be closed. The following examples can help the readers to understand these two hypotheses.

Example 2.6 (*Dependency on the domain topology: contractibility*) On the punctured plane $\mathbb{R}^2 \setminus \{(0, 0)\}$, consider the 1-form $\omega = d\theta$ where θ is the angle (the argument). It is known that $\int_{\mathbb{S}^1} d\theta = 2\pi$ where $\mathbb{S}^1$ represents the unit circle around the origin. Thus, the argument increases with 2π when we turn once around the origin on $\mathbb{S}^1$. In Cartesian coordinates (x, y) of the plan, this 1-form is expressed as follows:

$$\omega = \frac{1}{x^2 + y^2} (-ydx + xdy),$$

which is globally defined except at the origin. It is easy to show that $d\omega = 0$ but it is not a differential of a function because of the topology of $\mathbb{R}^2 \setminus \{(0, 0)\}$. Thus, the definition of the function h will depend on the path γ. □

Example 2.7 (*Dependency on the expression of* 1*-form: closure*) On the whole plan $\mathbb{R}^2$, consider the following 1-form $\omega = -ydx + xdy$. A straightforward calculation gives

$$d\omega = 2dx \wedge dy \neq 0,$$

where the $\wedge$ symbol denotes the exterior product that means the evaluation on two vector fields v and w is given by $dx \wedge dy(v, w) = dx(v)dy(w) - dx(w)dy(v)$. Therefore, it is not a differential of a function even if the domain is contractile. □

2.4 Change of Coordinates: Diffeomorphism

This subsection provides a method that enables us to find a change of coordinates that transforms (locally) a family of vector fields, which forms a basis, into a canonical basis of constant vector fields. This will be useful throughout this book.

Let $\mathscr{X}$ and $\mathscr{Z}$ be open neighborhoods of 0 in $\mathbb{R}^n$ and let the mapping [13]

$$\phi : \mathscr{X} \to \mathscr{Z}$$

be a change of coordinates (diffeomorphism). It is a well-known fact that the Jacobian ϕ_* of ϕ is an isomorphism on the tangent fiber bundles such that the following diagram commutes:

$$\begin{array}{ccc} T\mathscr{X} & \xrightarrow{\phi_*} & T\mathscr{Z} \\ p\downarrow & & \downarrow p \\ \mathscr{X} & \xrightarrow{\phi} & \mathscr{Z}. \end{array} \tag{2.17}$$

The Jacobian ϕ_* of $\phi = (\phi_1, \ldots, \phi_n)^T$ can be seen as n differential 1-forms $\omega = (\omega_1, \ldots, \omega_n)^T$ where $\{\omega_i = d\phi_i\}_{1 \le i \le n}$. These 1-forms are linearly independent because ϕ_* is an isomorphism.

The question about *"whether every family of n linearly independent 1-forms is a Jacobian of a change of coordinates"* has been solved by Poincaré's lemma 2.1. However, in this book, we will provide an algorithm to check the existence of such a change of coordinates and to compute it. Before stating this method, let us give an example to show how it works.

Example 2.8 In the plan of coordinates (x_1, x_2), consider the following two 1-forms:

$$\begin{aligned} \omega_1 &= dx_1 \\ \omega_2 &= x_2 dx_1 + dx_2 \end{aligned}$$

the first 1-form ω_1 is exact and the second one ω_2 is not, since

$$d\omega_2 = -dx_1 \wedge dx_2 \neq 0.$$

Now, let τ_1 and τ_2 be two vector fields that fulfill the following equations:

$$\omega_i(\tau_j) = \delta_i^j \text{ for } 1 \le i, j \le 2, \tag{2.18}$$

where $\delta_i^j = 1$ when $i = j$, otherwise it is equal to 0 for $i \neq j$. A simple verification shows that

$$\tau_1 = \frac{\partial}{\partial x_1} - x_2 \frac{\partial}{\partial x_2}$$
$$\tau_2 = \frac{\partial}{\partial x_2}$$

are the solutions of (2.18).

A straight calculation shows that the Lie bracket of these two vector fields is

$$[\tau_1, \tau_2] = \frac{\partial}{\partial x_2}.$$

Now, thanks to formula (2.16) we obtain

$$d\omega_2(\tau_1, \tau_2) = -\omega_2([\tau_1, \tau_2]) = -1 \neq 0.$$

As ω_1 and ω_2 are linearly independent, we can consider $\omega = (\omega_1, \omega_2)$ as an isomorphism of the tangent fiber bundle. However, it is not a differential of a change of coordinates because

$$1 = \omega([\tau_1, \tau_2]) \neq [\omega(\tau_1), \omega(\tau_2)] = 0.$$

Indeed, the two vector fields

$$\omega(\tau_1) = \begin{pmatrix} \omega_1(\tau_1) \\ \omega_2(\tau_1) \end{pmatrix} = \begin{pmatrix} 1 \\ 0 \end{pmatrix}$$

and

$$\omega(\tau_2) = \begin{pmatrix} \omega_1(\tau_2) \\ \omega_2(\tau_2) \end{pmatrix} = \begin{pmatrix} 0 \\ 1 \end{pmatrix}$$

are constant and therefore they commute while the vector fields τ_1 and τ_2 do not. □

Lemma 2.2 *Let $\tau = (\tau_i)_{1 \leq i \leq n}$ be a basis of the tangent fiber bundle $T\mathscr{X}$ and $\omega = (\omega_i)_{1 \leq i \leq n}$ be a basis of the cotangent fiber bundle $T^*\mathscr{X}$ such that the following equations are fulfilled:*

$$\omega_i(\tau_j) = \delta^i_j.$$

Thus ω is an isomorphism that transforms the basis τ into a constant basis of $\mathbb{R}^n$. Under the above assumption, the following statements are equivalent:

(1) There exists a change of coordinates (diffeomorphism) $z = \phi(x)$ such that

$$\omega = \phi_*.$$

(2) The Lie brackets satisfy

$$[\tau_i, \tau_j] = 0,$$

for all $1 \leq i, j \leq n$, *i.e., the vector fields* τ_i *and* τ_j *commute.*

(3) The 1*-forms* ω_i *for all* $1 \leq i \leq n$ *are closed, i.e.,* $d\omega_i = 0$*, or*

$$d\omega = 0.$$

In this case, the diffeomorphism is determined locally by

$$\phi_i(x) = z_i = \int_\gamma \omega_i(\gamma'(t))dt,$$

where $\gamma(t)$ *is any path linking* 0 *to* x. □

Proof Consider the formula (2.16) where we set $\theta = \omega_i$ and $f_j = \tau_j$. As $\omega_i(\tau_j) = \delta^i_j$ are constant, then formula (2.16) is equivalent to

$$d\omega_i(\tau_j, \tau_k) = -\omega_i[\tau_j, \tau_k].$$

As $\{\omega_i\}_{1\leq i\leq n}$ are linearly independent, then we obtain the equivalence between the second and the third statements of Lemma 2.2 . The equivalence between the first and the third statements is due to Poincaré's lemma 2.1. □

Example 2.9 In the plan of coordinates (x_1, x_2), consider the following two 1-forms:

$$\begin{aligned} \omega_1 &= dx_1 \\ \omega_2 &= x_1 dx_1 + dx_2 \end{aligned}$$

and the following two vector fields:

$$\begin{aligned} \tau_1 &= \frac{\partial}{\partial x_1} - x_1 \frac{\partial}{\partial x_2} \\ \tau_2 &= \frac{\partial}{\partial x_2}. \end{aligned}$$

A simple verification shows that these two 1-forms are closed and these two vector fields commute. Therefore, the desired change of coordinates is given by

$$\begin{aligned} z_1 &= x_1 \\ z_2 &= x_2 + \tfrac{1}{2}x_1^2. \end{aligned}$$

Therefore, $z = \phi(x_1, x_2) = (z_1, z_2)^T$ transforms the vector fields τ_1 and τ_2 to the constant vector fields $\frac{\partial}{\partial z_1}$ and $\frac{\partial}{\partial z_2}$ via its Jacobian $\phi_* = \omega$. □

2.5 Integrability, Involutivity and Frobenius Theorem

The notion of integrability of constraints in physical systems is an interesting engineering problem. Indeed, the behavior of certain systems depends on constraints due to their mechanical structures, as well as on the actuators that used to achieve specific tasks [11]. In many cases, these constraints are expressed by a set of algebraic equations. If these equations depend only on the coordinates, then they are called *holonomic constraints*. Holonomic constraints impose system's reachable configurations on a sub-manifold of the configuration space defined by these equations [5].

If these equations contain also time derivatives of the coordinates (i.e., the velocity) and it is not possible to obtain algebraic equations purely dependent on the coordinates by integration, then they are called *non-holonomic constraints*, which implies that those constraints are not integrable. In general, these kinds of constraints do not define a constraint sub-manifold of the whole configuration space.

An example of holonomic constraint (or integrable constraint) is given by a material point which is constrained to move on a circle of radius r in the plan $\mathbb{R}^2$ of coordinates (x, y). This constraint is expressed by

$$x^2 + y^2 - r = 0,$$

which is of the form $h(x, y) = 0$, depending only on the coordinates (x, y).

Hereafter, we will give two examples of non-holonomic constraints.

Example 2.10 Consider the motion of a wheel of radius R rolling without sliding on a horizontal plane (see Fig. 2.5).

Denote its configuration coordinates by $q = (x, y, \varphi, \beta)^T$, then its generalized velocity $\dot{q}$ fulfills the following constraints:

$$\omega_1(\dot{q}) = 0$$
$$\omega_2(\dot{q}) = 0,$$

where ω_1 and ω_2 are the two 1-forms defined by

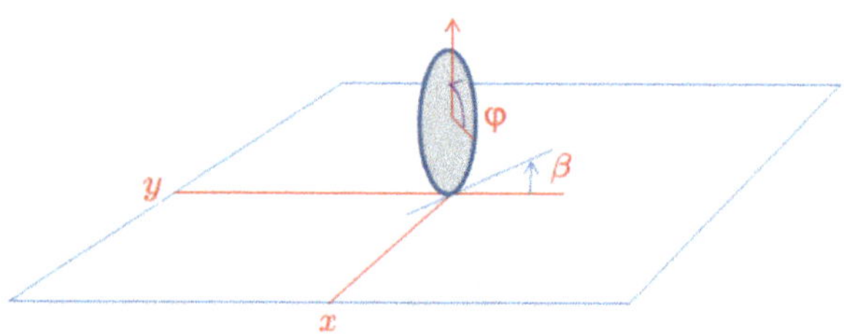

Fig. 2.5 Non-holonomic wheel

$$\omega_1 = dx - R\cos\beta d\varphi$$
$$\omega_2 = dy - R\sin\beta d\varphi.$$

It is easy to check that these two 1-forms are not closed, and therefore they cannot be the differential of a function of the coordinates q. However, this argument is not sufficient to deduce that these two 1-forms do not define a constraint sub-manifold, i.e., they are not integrable.

To give an idea why there is no such a sub-manifold, it is worth noting that the directions in which the wheel moves are in the kernel of these two 1-forms. This kernel is a two-dimensional distribution spanned by

$$X_1 = R\cos\beta\frac{\partial}{\partial x} + R\sin\beta\frac{\partial}{\partial y} + \frac{\partial}{\partial\varphi}$$
$$X_2 = \frac{\partial}{\partial\beta}.$$

As the wheel evolves in X_1 and X_2 directions, then it also evolves in the direction $X_3 = [X_2, X_1]$ provided by their Lie bracket. In fact, this direction is obtained by driving the wheel in the direction X_1, then X_2 and $-X_1$, and finally in the direction $-X_2$. Thus, the two directions X_1 and X_2 give rise to the following new direction

$$X_3 = [X_2, X_1] = -R\sin\beta\frac{\partial}{\partial x} + R\cos\beta\frac{\partial}{\partial y},$$

which is independent of X_1 and X_2. Once again, the Lie bracket of X_2 with this new direction X_3 gives rise to the fourth independent direction

$$X_4 = -R\cos\beta\frac{\partial}{\partial x} - R\sin\beta\frac{\partial}{\partial y}.$$

Therefore, in the four-dimensional space of configuration, with coordinates $q = (x, y, \varphi, \beta)^T$, the system can evolve in four independent directions. Thus, thanks to these four directions the wheel can reach any configuration $q = (x, y, \varphi, \beta)^T$. In summary, even if the wheel should satisfy the non-holonomic constraints, it is not constrained in space configuration. □

Example 2.11 Consider the well-known car-like robot (see Fig. 2.6) endowed with the generalized coordinates $q = (x, y, \theta, \phi)^T$, where (x, y) are the Cartesian coordinates of the rear-axle midpoint, θ the orientation of the car body and ϕ represents the steering angle.

This system evolves under the following two non-holonomic constraints:

$$\omega_1(\dot{q}) = 0$$
$$\omega_2(\dot{q}) = 0,$$

Fig. 2.6 Non-holonomic car-like robot

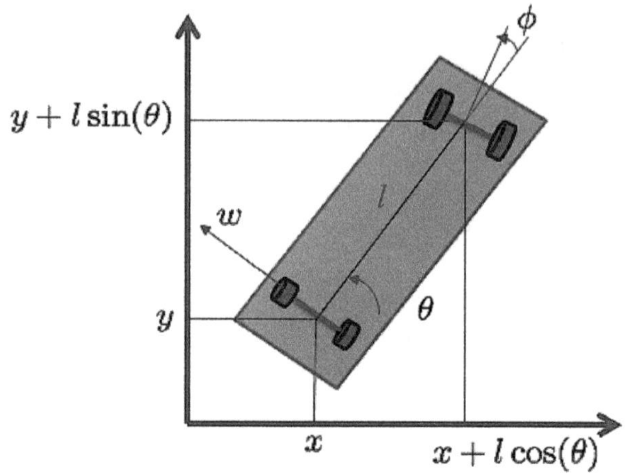

where ω_1 and ω_2 are the following two 1-forms:

$$\begin{aligned}\omega_1 &= \sin\theta dx - \cos\theta dy \\ \omega_2 &= \sin(\theta+\phi)dx - \cos(\theta+\phi)dy - l\cos\phi d\theta.\end{aligned}$$

A straightforward calculation shows that the kernel of these two 1-forms (the space of admissible generalized velocities) is the distribution spanned by

$$\begin{aligned}X_1 &= \cos\theta\cos\phi\frac{\partial}{\partial x} + \sin\theta\cos\phi\frac{\partial}{\partial y} + \frac{1}{l}\sin\phi\frac{\partial}{\partial\theta} \\ X_2 &= \frac{\partial}{\partial\phi}.\end{aligned}$$

As for the above example, these two directions can generate the following new direction:

$$X_3 = [X_2, X_1] = -\cos\theta\sin\phi\frac{\partial}{\partial x} - \sin\theta\sin\phi\frac{\partial}{\partial y} + \frac{1}{l}\cos\phi\frac{\partial}{\partial\theta},$$

which is independent of X_1 and X_2. In the same way, this new direction, together with X_2, generates as well the following direction:

$$X_4 = [X_2, X_3] = -\cos\theta\cos\phi\frac{\partial}{\partial x} - \sin\theta\cos\phi\frac{\partial}{\partial y} - \frac{1}{l}\sin\phi\frac{\partial}{\partial\theta},$$

which is the same as $-X_1$. However, the Lie bracket of X_1 and X_3 gives rise to the following independent direction:

$$X_5 = [X_1, X_3] = \frac{1}{l}\left(\sin\theta\frac{\partial}{\partial x} - \cos\theta\frac{\partial}{\partial y}\right).$$

Finally, since the configuration system of the car-like robot has four possible independent directions in a four-dimensional space, it is not constrained. □

Integrability of a set of 1-forms does not mean that each of them is a differential of a function. To clarify the true meaning of the integrability, let us denote

$$\Delta^{\perp} = \text{span}\{\omega_1, \ldots, \omega_{n-m}\}$$

as the co-distribution spanned by $n - m$ linearly independent 1-forms on a manifold $\mathbb{R}^n$. The distribution

$$\Delta = \text{span}\{X_1, \ldots, X_m\}$$

is spanned by m linearly independent vector fields belonging to the kernel of each 1-form ω_i for $1 \leq i \leq n - m$. Assume that the dimension of Δ is equal to m everywhere in its domain. In this case, we say that Δ is a regular distribution, and therefore the same is true for the co-distribution $\Delta^{\perp}$ which is of dimension $n - r$. Hereafter, we state the well-known Frobenius theorem.

Lemma 2.3 *The distribution Δ (or co-distribution $\Delta^{\perp}$) is involutive (or integrable) if and only if one of the following equivalent statements holds:*

(1) The Lie bracket $[X_i, X_j] \in \Delta$ for all $1 \leq i, j \leq m$, i.e., the distribution Δ is Lie bracket closed.
(2) The differential $d\theta_i = \sum_{j=1}^{n-m} \omega_{i,j} \wedge \theta_j$ for all $1 \leq i \leq n - m$ where $\wedge$ is the wedge product and $\omega_{i,j}$ are 1-forms.

The above result is well known in foliation theory (generalization of integral curve of a vector field), and the interested reader can refer to [1] for the proof. In general, Lemma 2.3 states that, if a distribution Δ is involutive, then it is tangent to a foliation [3, 6]. Roughly speaking, a foliation splits a manifold into disjoint connected sub-manifolds of same dimension, called leaves. This theory was initiated by [9, 14] in the 40s of the last century. When the initial conditions are fixed, then the leaf that contains them becomes a constraint sub-manifold, in which the configuration coordinates of motion of a system are laid. Thus, without initial conditions integrable kinematic constraint equations define many constraint sub-manifolds: leaves.

The following examples highlight the above results.

Example 2.12 Reeb foliation: In the plan $\mathbb{R}^2$ of coordinates (x_1, x_2), consider the co-distribution $\Delta^{\perp}$ spanned by the 1-form

$$\omega = 2x_1 dx_1 + (x_1^2 - 1)dx_2.$$

It is easy to check that the corresponding distribution Δ is spanned by the vector field

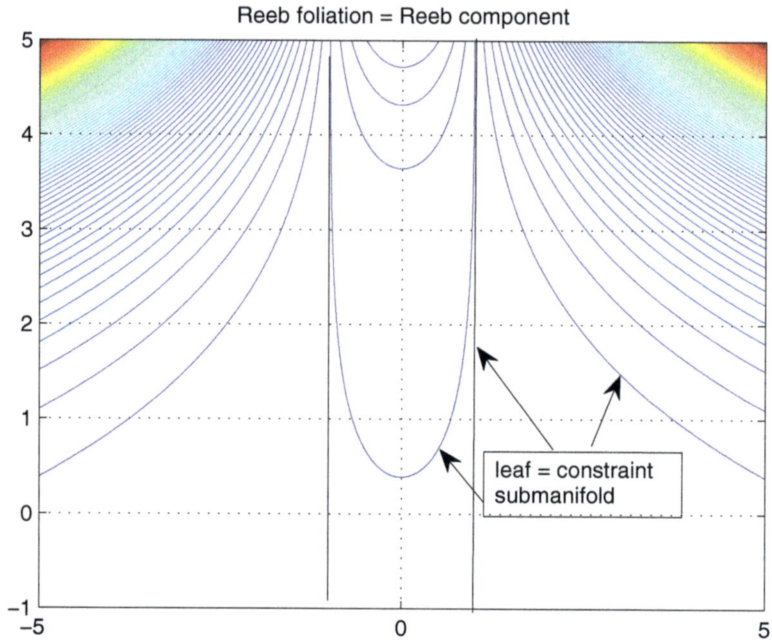

Fig. 2.7 Reeb foliation in the plan

$$X = (x_1^2 - 1)\frac{\partial}{\partial x_1} - 2x_1\frac{\partial}{\partial x_2}.$$

Since there exists only one vector field, Δ is always Lie bracket closed, i.e., the first item of Lemma 2.3 is fulfilled, which implies that Δ is involutive.

We can also check the equivalent second item of Lemma 2.3. The straightforward calculation gives the differential

$$d\omega = 2x_1 dx_1 \wedge dx_2 = -dx_2 \wedge \omega,$$

thus the second item of Lemma 2.3 is fulfilled. Consequently, $\Delta^{\perp}$ generates a foliation given by the level curves of the function $h(x_1, x_2) = (x_1^2 - 1)e^{x_2}$ (see Fig. 2.7). In fact, the differential of this function is as follows:

$$dh(x_1, x_2) = e^{x_2}(2x_1 dx_1 + (x_1^2 - 1)dx_2) = e^{x_2}\omega \in \Delta^{\perp},$$

and therefore it spans the same co-distribution as ω. □

Example 2.13 Consider the following 1-forms:

$$\omega_1 = dx_1 + x_1 dx_2$$
$$\omega_2 = dx_2 + x_2 dx_3,$$

thus their kernel is spanned by the vector field

$$X = x_2 x_1 \frac{\partial}{\partial x_1} - x_2 \frac{\partial}{\partial x_2} + \frac{\partial}{\partial x_3}.$$

It can be shown that $d\omega_1 = -dx_2 \wedge \omega_1$ and $d\omega_2 = -dx_3 \wedge \omega_2$. Thus, these 1-forms fulfill the second item of Lemma 2.3. Moreover, it can be proven as well that the vector field X fulfills the first equivalent item of Lemma 2.3. Thus, the distribution

$$\Delta = \ker\{\omega_1, \omega_2\} = \operatorname{span}\{X\}$$

is integrable. Therefore, Δ defines a foliation and each leaf represents a constraint sub-manifold. It can be shown that $e^{x_2}\omega_1 = d(x_1 e^{x_2})$ and $e^{x_3}\omega_1 = d(x_2 e^{x_3})$. Thus, the leaves of the co-distribution $\Delta^{\perp} = \operatorname{span}\{\omega_1, \omega_2\}$ are the level curve on $\mathbb{R}^3$. □

The following results are important if the distribution is involutive.

Theorem 2.1 ([12]) *If a distribution Δ of dimension r is involutive, then it admits a commutative basis. Moreover, there exist locally two families of vector fields $\{X_1, \ldots, X_r\}$ and $\{Y_1, \ldots, Y_{n-r}\}$ such that*

- *these two families of vector fields span the whole vector bundle;*
- *the family $\{X_1, \ldots, X_r\}$ spans the distribution Δ;*
- *$[Z_i, Z_j] = 0$ where Z_i and Z_j are any vector fields in the union of these two families.*

Definition 2.3 Let f be a vector field. Then, a distribution Δ (or co-distribution $\Delta^{\perp}$) is said to be f-invariant if and only if one of the following equivalent items is fulfilled:

- $[f, X] \in \Delta$ for any $X \in \Delta$;
- $L_f \omega \in \Delta^{\perp}$ for any $\omega \in \Delta^{\perp}$.

Moreover, if the distribution Δ is involutive, then the defined foliation is called to be f-invariant. Thus, the flow of f sends leaf to leaf.

Hereafter, we state a useful lemma that enables us to split the dynamics into observable/unobservable part and controllable/uncontrollable part.

Lemma 2.4 *If the distribution Δ is involutive and f-invariant, then locally the vector field f is split as follows:*

$$f = \sum_{i=1}^{r} a_i X_i + \sum_{j=1}^{n-r} b_j Y_j \text{ with } L_{X_i} b_j = 0,$$

and this implies that the functions b_j in the direction of X_i do not depend on the leaf coordinates.

Proof As X_i and Y_j form a basis, then $f = \sum_{i=1}^{r} a_i X_i + \sum_{j=1}^{n-r} b_j Y_j$. In order to prove the above lemma, we compute the Lie bracket of X_k for $1 \leq k \leq r$ with f, which yields

$$[X_k, f] = \sum_{i=1}^{r} (L_{X_k} a_i) X_i + \sum_{j=1}^{n-r} (L_{X_k} b_j) Y_j,$$

where we applied the facts that $[X_k, X_i] = 0$ and $[X_k, Y_j] = 0$. As the distribution spanned by X_i is f-invariant, we have $[X_k, f] \in \Delta$. Finally, we have $L_{X_k} b_j = 0$. □

Example 2.14 Consider again the Reeb foliation given by the following 1-form:

$$\omega = 2x_1 dx_1 + (x_1^2 - 1) dx_2$$

or by the following vector field:

$$X = (x_1^2 - 1) \frac{\partial}{\partial x_1} - x_1 \frac{\partial}{\partial x_2}.$$

Clearly, the vector field $Y = \frac{\partial}{\partial x_2}$, together with the vector field X, forms a basis. Moreover, we have $[X, Y] = 0$.

Exercises

Exercise 2.1 Consider the following 1-forms

$$\omega_1 = dx_1 + x_1 dx_2 + dx_3$$

and

$$\omega_2 = dx_1 + dx_2 + x_1 dx_3.$$

(1) Show that their kernel is spanned by the following vector field:

$$X = -(1 + x_1) \frac{\partial}{\partial x_1} + \frac{\partial}{\partial x_2} + \frac{\partial}{\partial x_3}.$$

(2) Show that they have the same kernel as the differential of the following two linearly independent functions:

$$h_1(x_1, x_2, x_3) = \ln(1 + x_1) + x_2$$

and

$$h_2(x_1, x_2, x_3) = x_2 - x_3.$$

(3) Show that they are integrable, thus they define constraint sub-manifolds.

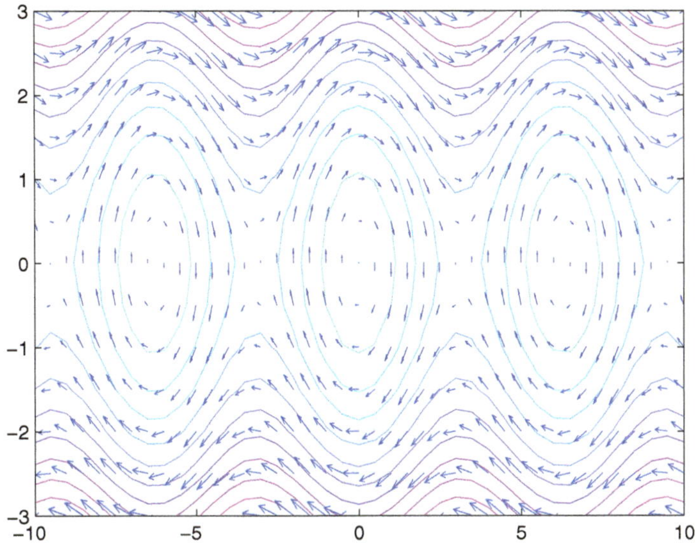

Fig. 2.8 Phase portrait of the system's motion

Exercise 2.2 Consider the well-known van der Pol oscillator with nonlinear damping described by the following equation:

$$\ddot{x} - \mu(1 - x^2)\dot{x} + x = 0, \tag{2.19}$$

and note $E = \frac{\dot{x}^2}{2} + \frac{x^2}{2}$ as the energy of the above system.

(1) Write (2.19) in the state-space representation as $\dot{X} = f(X)$ with $X = [x, \dot{x}]^T$.
(2) If $\mu \neq 0$, show that this system is a non-conservative oscillator, which means that $L_f E \neq 0$, where f is the vector field obtained in the first question.
(3) If $\mu = 0$, show that $L_f E = 0$ and verify that the phase portrait of this system can be depicted as Fig. 2.8.

Exercise 2.3 (1) Consider the Duffing oscillator given by the following equation:

$$\ddot{x} + \delta\dot{x} + \alpha x + \beta x^3 = 0,$$

where δ is the damping coefficient. Answer the same questions (1)–(3) of Exercise 2.2.

(2) In the plane (x, t) where x denotes a variable space and $t \geq 0$ denotes the time. Let the vector field $T = c\frac{\partial}{\partial x} + \frac{\partial}{\partial t}$ and a function $u(x, t)$ that fulfill the so-called transport equation given as

$$\begin{cases} Tu = c\frac{\partial u}{\partial x} + \frac{\partial u}{\partial t} = 0 \\ u(x, 0) = \varphi(x), \end{cases} \tag{2.20}$$

thus $u(x,t)$ is a constant on lines of the slope equal to $\frac{1}{c}$. Show that u is constant on the curve $x(t) = ct + x_0$, and we have

$$u(ct + x_0, t) = u(x_0, 0) = u(x - ct, 0) = \varphi(x - ct).$$

Exercise 2.4 Consider the following 1-form:

$$\omega = (1 - 2x_1^2)dx_1 - 2x_1x_2dx_2.$$

(1) Is there any function h such that $dh = \omega$?
(2) Consider the following vector field:

$$f_1 = 2x_1x_2\frac{\partial}{\partial x_1} + (1 - 2x_1^2)\frac{\partial}{\partial x_2}.$$

Compute $\omega(f_1)$, then give a vector field that spans the kernel of ω.
(3) Let

$$\omega_1 = e^{-(x_1^2+x_2^2)}\omega.$$

Find a function $h(x_1, x_2)$ such that $dh = \omega_1$. What is the relationship between the level curves of h and the integral curves of f_1?
(4) Consider the following vector field:

$$f_2 = -e^{-(x_1^2+x_2^2)}\left((1 - 2x_1^2)\frac{\partial}{\partial x_1} - 2x_1x_2\frac{\partial}{\partial x_2}\right).$$

Show that $f_2 = -\nabla h$ where ∇h is the gradient of h.
(5) Let $V = h$ and compute $L_{f_2}V$ to show that the singularity $(-\frac{1}{\sqrt{2}}, 0)$ of f_2 is asymptotically stable. How about its second singularity $(\frac{1}{\sqrt{2}}, 0)$?
(6) Are the singularities of f_1 stable, asymptotically stable or unstable?

Exercise 2.5 (1) Find the state-space representation of the following differential equation:

$$\ddot{y} + \delta\dot{y} + \alpha y + \beta y^3 = \gamma\cos(\omega t). \tag{2.21}$$

(2) Consider the following vector field:

$$f(x) = x_2\frac{\partial}{\partial x_1} - \left(\alpha x_1 + \delta x_2 + \beta x_1^3\right)\frac{\partial}{\partial x_2}, \tag{2.22}$$

what is the link between this vector field and the differential equation (2.21)?
(3) Find the singularity of the following dynamics:

$$\dot{x} = f(x),$$

where f is defined in (2.22). Study the stability of those singularities.

(4) Set $\beta = 0$ for $\dot{x} = f(x)$ where f is given in (2.22) and find the matrix A such that $f(x) = Ax$. Give the Laplace transformation of the linear dynamical system $\dot{x} = Ax$. Discuss the solution of this linear dynamical system based on the values of α and δ.

(5) Consider the following function:

$$V(x_1, x_2) = \frac{1}{2}x_2^2 + \frac{1}{2}\alpha x_1^2 + \beta\frac{1}{4}x_1^4.$$

- Compute the derivative of V: $\dot{V} = L_f V$.
- Show that the set where $L_f V = 0$ is given by the equation $x_2 = 0$.
- Prove that the f is transverse to the line $x_2 = 0$ if $x_1 \neq 0$.
- Deduce that for $\delta > 0$, $(0, 0)$ is asymptotically stable.

Exercise 2.6 Consider the following vector field:

$$f = -x_2\frac{\partial}{\partial x_1} + x_1\frac{\partial}{\partial x_2}$$

on $\mathbb{R}^2$.

(1) Draw the integral curves of f.

(2) Let

$$\omega_1 = \frac{-x_2}{x_1^2 + x_2^2}dx_1 + \frac{x_1}{x_1^2 + x_2^2}dx_1$$

and

$$\omega_2 = x_1 dx_2 + x_2 dx_1$$

on $\mathbb{R}^2\backslash\{(0, 0)\}$.

- Show that $\omega = \begin{pmatrix}\omega_1\\ \omega_2\end{pmatrix}$ is an isomorphism on the tangent space of $\mathbb{R}^2\backslash\{(0, 0)\}$.
- Prove that $\omega(f) = \begin{pmatrix}1\\ 0\end{pmatrix}$ is a constant vector field.
- Let $g = x_1\frac{\partial}{\partial x_1} + x_2\frac{\partial}{\partial x_2}$, compute $\omega(g)$ and deduce that there does not exist a change of coordinates $\phi(x) = z$ such that $\omega = d\phi$.
- Could we judge directly from the expressions ω_1 and ω_2 that there does not exist a change of coordinates $\phi(x) = z$ such that $\omega = d\phi$?

References

1. Abraham, R., Marsden, J., Marsden, J.: Foundations of Mechanics, vol. 36. Benjamin/Cummings Publishing Company, Reading (1978)
2. Burke, W.L.: Applied Differential Geometry. Cambridge University Press, Cambridge (1985)

3. Camacho, C., Neto, A.L.: Geometric Theory of Foliations. Springer Science & Business Media, New York (2013)
4. Chen, W., Chern, S.S., Lam, K.S.: Lectures on Differential Geometry, vol. 1. World Scientific Publishing Company, Singapore (1999)
5. Crainic, M., Fernandes, R.L.: Integrability of Lie brackets. Ann. Math. 575–620 (2003)
6. Hector, G., Hirsch, U.: Introduction to the Geometry of Foliations. Springer, Berlin (1981)
7. Hicks, N.J.: Notes on Differential Geometry, vol. 3. van Nostrand, Princeton (1965)
8. Lang, S.: Fundamentals of Differential Geometry, vol. 191. Springer Science & Business Media, New York (2012)
9. Matsumoto, S.: Measure of exceptional minimal sets of codimension one foliations. A Fête of Topology, pp. 81–94. Elsevier, Amsterdam (1988)
10. Morita, S.: Geometry of Differential Forms, vol. 201. American Mathematical Society, Providence (2001)
11. Murray, R.M., Li, Z., Sastry, S.S., Sastry, S.S.: A Mathematical Introduction to Robotic Manipulation. CRC Press, Boca Raton (1994)
12. Nijmeijer, H., Van der Schaft, A.: Nonlinear Dynamical Control Systems, vol. 175. Springer, Berlin (1990)
13. Sternberg, S.: Local contractions and a theorem of Poincaré. Am. J. Math. **79**(4), 809–824 (1957)
14. Wolak, R.: Ehresmann connections for Lagrangian foliations. J. Geom. Phys. **17**(4), 310–320 (1995)

Chapter 3
Observer Normal Form with Output Injection

Abstract This chapter firstly presents the pioneering work of Krener [6, 7], for the purpose of transforming a general nonlinear system into the so-called nonlinear observer normal form with output injection. As it has been already mentioned in Chap. 1, this form contains a linear part which is of Brunovsky's canonical form, and a nonlinear part which is only function of measurable variables (i.e., input and output). A set of geometric conditions have been deduced to guarantee the existence of such a change of coordinates (a diffeomorphism) which transforms the studied nonlinear dynamical system into the proposed observer normal form with output injection. In addition, a constructive method has been proposed to facilitate the deduction of this diffeomorphism. After that, we consider how to extend such a method to treat nonlinear systems with inputs, for which the additional Lie bracket conditions need to be considered. Due to the special form of this observer normal form with output injection, a simple Luenberger-like observer [8] can be designed which yields a linear dynamics for the observation error. The concrete design procedures for this type of observer will be discussed in the last section of this chapter.

3.1 Problem Statement

Consider the following single-output dynamical system:

$$\begin{aligned} \dot{x} &= f(x) \\ y &= h(x), \end{aligned} \tag{3.1}$$

where $\mathscr{X} \subseteq \mathbb{R}^n$ is a neighborhood of 0, $x \in \mathscr{X}$ and $y \in \mathbb{R}$. Without loss of generality it is assumed that f and h are smooth, $f(0) = 0$ and $h(0) = 0$. By using the notations in Chap. 2, the vector field f can be written as

$$f(x) = \sum_{i=1}^{n} f_i(x) \frac{\partial}{\partial x_i}.$$

D. Boutat and G. Zheng, *Observer Design for Nonlinear Dynamical Systems*,
Lecture Notes in Control and Information Sciences 487,
https://doi.org/10.1007/978-3-030-73742-9_3

For the given single-output dynamical system (3.1), we can define the so-called observability 1-forms.

Definition 3.1 The observability 1-forms of dynamical system (3.1) are defined as follows:

$$\theta_i = dL_f^{i-1}h, \text{ for } 1 \leq i \leq n, \tag{3.2}$$

where d denotes the differential and L_f^k is the kth Lie derivative in the direction of the vector field f.

Within this chapter, it is assumed that dynamical system (3.1) fulfills the observability rank condition in the sense that its observability 1-forms θ_i for $1 \leq i \leq n$ are linearly independent. In other words, the whole state of (3.1) is assumed to be observable. And we are interested in how to seek a local diffeomorphism $z = \phi(x)$ such that the studied dynamical system (3.1) can be transformed into the following nonlinear observer normal form with output injection:

$$\begin{aligned} \dot{z} &= A_O z + \beta(y) \\ y &= C_O z = z_n, \end{aligned} \tag{3.3}$$

where A_O and C_O are of Brunovsky form, i.e.,

$$A_O = \begin{pmatrix} 0 & \cdots & 0 & 0 & 0 \\ 1 & \cdots & 0 & 0 & 0 \\ \vdots & \ddots & \ddots & \vdots & \vdots \\ 0 & \cdots & 1 & 0 & 0 \\ 0 & \cdots & 0 & 1 & 0 \end{pmatrix} \in \mathbb{R}^{n \times n}$$

and

$$C_O = \begin{pmatrix} 0 & \cdots & 0 & 1 \end{pmatrix} \in \mathbb{R}^{1 \times n}.$$

This normal form can be seen as a nonlinear version of (1.18), since it enables us to easily design a Luenberger-like observer to estimate its state (see Sect. 3.4 for the details). Such a problem was firstly investigated in [6] where necessary and sufficient conditions were stated to guarantee the existence of such a change of coordinates $z = \phi(x)$. However, no constructive approach was presented to calculate such a $\phi(x)$. This chapter aims at revisiting the result stated in [6], and presents the explicit formulas to compute such a diffeomorphism.

3.2 Observer Normal Form with Output Injection

For the dynamical system (3.1), we can define a frame $\tau = [\tau_1, \ldots, \tau_n]$ where τ_i for $1 \leq i \leq n$ are vector fields defined according to the following criteria:

- the first vector field τ_1 is the unique solution of the following algebraic equations:

$$\begin{aligned} \theta_k(\tau_1) &= 0, \text{ for } 1 \leq k \leq n-1 \\ \theta_n(\tau_1) &= 1; \end{aligned} \tag{3.4}$$

- the rest vector fields are calculated via the Lie bracket by the following induction:

$$\tau_i = \left[\tau_{i-1}, f\right], \text{ for } 2 \leq i \leq n. \tag{3.5}$$

The following recalls the result stated in [4, 6].

Theorem 3.1 *There exists a local diffeomorphism $z = \phi(x)$ that transforms system (3.1) into the nonlinear observer normal form (3.3) if and only if*

$$\left[\tau_i, \tau_j\right] = 0, \tag{3.6}$$

where τ_i and τ_j for $1 \leq i, j \leq n$ are defined in (3.4)–(3.5).

The proof of the above theorem can be found in [5].

Theorem 3.1 presents only the conditions on the existence of such a diffeomorphism, but does not provide a constructive way to calculate it. In what follows, we will present an explicit formula to compute such a diffeomorphism.

For this, let us consider the following matrix:

$$\Lambda = \theta\tau = \left(\theta_i\left(\tau_j\right)\right)_{1\leq i,j\leq n} = \left(\Lambda_{i,j}\right)_{1\leq i,j\leq n}, \tag{3.7}$$

where

$$\theta = \begin{bmatrix} \theta_1 \\ \vdots \\ \theta_n \end{bmatrix}, \tau = [\tau_1, \ldots, \tau_n].$$

The following proposition gives the algorithm to compute Λ's expression.

Proposition 3.1 *The coefficients of the matrix Λ, defined in (3.7), satisfy the following relation:*

$$\Lambda_{i,j+1} = \Lambda_{i+1,j} - L_f\left(\Lambda_{i,j}\right). \tag{3.8}$$

Particularly, if $\Lambda_{i,j}$ is a constant function, then $\Lambda_{i,j+1} = \Lambda_{i+1,j}$.

Proof Recall that

$$d\theta_i\left(\tau_j, f\right) = L_{\tau_j}\theta_i(f) - L_f\theta_i\left(\tau_j\right) - \theta_i\left(\left[\tau_j, f\right]\right). \tag{3.9}$$

As θ_i are exact differential 1-forms, then $d\theta_i = 0$. Moreover, for $1 \leq j \leq n-1$, by definition we know that $\tau_{j+1} = [\tau_j, f]$. Therefore, Eq. (3.9) becomes

$$\theta_i\left(\tau_{j+1}\right) = L_{\tau_j}\theta_i(f) - L_f\theta_i\left(\tau_j\right).$$

Recall that the Lie derivative can be written as $L = d\iota + \iota d$ where ι is the antiderivation operator (or interior product) and d the differential. Then, we deduce that

$$\theta_i\left(\tau_{j+1}\right) = \theta_{i+1}\left(\tau_j\right) - L_f\left(\theta_i\left(\tau_j\right)\right).$$

Thus, according to the definition of $\Lambda_{i,j}$, the above equality can be written as that in (3.8).

Moreover, if $\theta_i\left(\tau_j\right)$ is a constant function, then its Lie derivative vanishes and we have $\theta_i\left(\tau_{j+1}\right) = \theta_{i+1}\left(\tau_j\right)$, i.e., $\Lambda_{i,j+1} = \Lambda_{i+1,j}$. □

Following the above result, we know that $\theta_i(\tau_1) = 0$ for $1 \le i \le n-1$ and $\theta_n(\tau_1) = 1$, then we deduce that $\theta_i(\tau_2) = 0$ for $1 \le i \le n-2$ and $\theta_{n-1}(\tau_2) = 1$. By repeating the same procedure, a straightforward calculation shows that Λ has the following form:

$$\Lambda = \begin{pmatrix} 0 & 0 & \cdots & 0 & 1 \\ 0 & 0 & \cdots & 1 & \Lambda_{2,n} \\ \vdots & \vdots & \cdot^{\cdot^{\cdot}} & \vdots & \vdots \\ 0 & 1 & \cdots & \Lambda_{n-1,n-1} & \Lambda_{n-1,n} \\ 1 & \Lambda_{n,2} & \cdots & \Lambda_{n,n-1} & \Lambda_{n,n} \end{pmatrix}. \tag{3.10}$$

It is clear that this matrix is invertible, thus we can define the following 1-forms:

$$\omega = \Lambda^{-1}\theta := (\omega_i)_{1\le i\le n}, \tag{3.11}$$

where $(\omega_i)_{1\le i\le n}$ are given by the following equations [1, 2]:

$$\begin{aligned} \omega_n &= \theta_1 \\ \omega_{n-r} &= \left(\theta_{r+1} - \sum_{i=n-r+1}^{n} \Lambda_{r+1,i}\omega_i\right), \text{ for } 1 \le r \le n-1. \end{aligned} \tag{3.12}$$

Theorem 3.2 *The following three assertions are equivalent [3]:*

(1) There exists a local change of coordinates $z = \phi(x)$ that transforms system (3.1) into the nonlinear observer normal form (3.3);
(2) The vector fields τ_i and τ_j for $1 \le i, j \le n$ defined in (3.4)–(3.5) satisfy the commutativity condition (3.6);
(3) The 1-forms ω_i defined in (3.12) for $1 \le i \le n$ are closed, i.e.,

$$d\omega_i = 0. \tag{3.13}$$

Furthermore, the change of coordinates $z = \phi(x)$ is determined by

$$z_i = \phi_i(x) = \int_\gamma \omega_i, \tag{3.14}$$

where $\gamma : [0, 1] \to \mathcal{X}$ is any smooth curve on contractile neighborhood of 0 *such that* $\gamma(0) = 0$ *and* $\gamma(1) = x$.

Proof The equivalence between (3.6) in the second assertion and (3.13) in the third assertion was proven in Lemma 2.2 of Sect. 2.4. The rest is to prove the equivalence between the first assertion and the second one.

For this, let ϕ be the diffeomorphism given in (3.14), which fulfills $\phi_* := d\phi = \omega$. The following shows how the Jacobian of this diffeomorphism ϕ transforms the vector field f, which is, in fact, equivalent to check $\frac{\partial}{\partial z_i}\phi_*(f)$.

Using Eq. (2.16) with $f_1 = \tau_i$ and $f_2 = f$, we obtain

$$\frac{\partial}{\partial z_i}\phi_*(f) = \phi_*[\tau_i, f] = \phi_*(\tau_{i+1}) = \frac{\partial}{\partial z_{i+1}}$$

for all $1 \le i \le n-1$. The integration of the above equation leads to

$$\dot{z} = \phi_*(f) = A_O z + \beta(z_n),$$

where z_n is determined by (3.14) as

$$z_n = \int_\gamma \omega_n.$$

Based on (3.12) and $\theta_1 = dh = dy$, we finally have $y = z_n$. Therefore, we proved that the diffeomorphism $z = \phi(x)$ transforms dynamical system (3.1) into the following observer normal form:

$$\begin{aligned} \dot{z} &= \phi_*(f) = A_O z + \beta(z_n) \\ y &= z_n. \end{aligned}$$

□

In order to clearly understand the above result, let us firstly apply Theorem 3.2 to linear dynamical systems.

Example 3.1 Consider the following linear dynamical system:

$$\begin{aligned} \dot{x} &= Ax \\ y &= Cx \end{aligned}$$

with

$$A = \begin{pmatrix} 1 & -1 & 0 \\ 1 & 2 & 0 \\ 0 & 1 & 1 \end{pmatrix} \text{ and } C = \begin{pmatrix} 0 & 1 & -1 \end{pmatrix}.$$

Let us apply the proposed procedure to calculate the observability 1-forms θ, defined in (3.2), and this yields

$$\begin{aligned}\theta_1 &= C = \begin{pmatrix}0 & 1 & -1\end{pmatrix} = dx_2 - dx_3 \\ \theta_2 &= CA = \begin{pmatrix}1 & 1 & -1\end{pmatrix} = dx_1 + dx_2 - dx_3 \\ \theta_3 &= CA^2 = \begin{pmatrix}2 & 0 & -1\end{pmatrix} = 2dx_1 - dx_3.\end{aligned}$$

Obviously, these 1-forms are linearly independent, thus the dynamical system fulfills the observability rank condition. Then, we need to calculate the frame $\tau = [\tau_1, \tau_2, \tau_3]$. According to (3.4), the first vector field τ_1 is the unique solution of the following algebraic equations:

$$\begin{aligned}C\tau_1 &= 0 \\ CA\tau_1 &= 0 \\ CA^2\tau_1 &= 1.\end{aligned}$$

A straightforward calculation gives

$$\tau_1 = \begin{pmatrix}0 \\ -1 \\ -1\end{pmatrix} = -\frac{\partial}{\partial x_2} - \frac{\partial}{\partial x_3}.$$

Then, according to (3.5), the second and the third vector fields are

$$\tau_2 = [\tau_1, Ax] = A\tau_1 = \begin{pmatrix}1 \\ -2 \\ -2\end{pmatrix} = \frac{\partial}{\partial x_1} - 2\frac{\partial}{\partial x_2} - 2\frac{\partial}{\partial x_3}$$

and

$$\tau_3 = [\tau_2, Ax] = A\tau_2 = \begin{pmatrix}3 \\ -3 \\ -4\end{pmatrix} = 3\frac{\partial}{\partial x_1} - 3\frac{\partial}{\partial x_2} - 4\frac{\partial}{\partial x_3}.$$

According to (3.7), we can then calculate Λ as

$$\Lambda = \theta\tau = \begin{pmatrix}C\tau_1 & C\tau_2 & C\tau_3 \\ CA\tau_1 & CA\tau_2 & CA\tau_3 \\ CA^2\tau_1 & CA^2\tau_2 & CA^2\tau_3\end{pmatrix} = \begin{pmatrix}0 & 0 & 1 \\ 0 & 1 & 4 \\ 1 & 4 & 10\end{pmatrix},$$

which yields

$$\Lambda^{-1} = \begin{pmatrix}6 & -4 & 1 \\ -4 & 1 & 0 \\ 1 & 0 & 0\end{pmatrix}.$$

Thus

$$\omega = \Lambda^{-1}\theta = \Lambda^{-1}\begin{pmatrix} C \\ CA \\ CA^2 \end{pmatrix} = \begin{pmatrix} -2 & 2 & -3 \\ 1 & -3 & 3 \\ 0 & 1 & -1 \end{pmatrix} = \begin{pmatrix} -2dx_1 + 2dx_2 - 3dx_3 \\ dx_1 - 3dx_2 + 3dx_3 \\ dx_2 - dx_3 \end{pmatrix}.$$

As a result, by integration of ω, the following diffeomorphism is obtained:

$$z = \begin{pmatrix} z_1 \\ z_2 \\ z_3 \end{pmatrix} = \begin{pmatrix} -2x_1 + 2x_2 - 3x_3 \\ x_1 - 3x_2 + 3x_3 \\ x_2 - x_3 \end{pmatrix},$$

which transforms the studied example into the following observer normal form:

$$\begin{aligned} \dot{z}_1 &= \beta_1(y) \\ \dot{z}_2 &= z_1 + \beta_2(y) \\ \dot{z}_3 &= z_2 + \beta_3(y) \\ y &= z_3, \end{aligned}$$

with $\beta_1(y) = 3y$, $\beta_2(y) = -6y$ and $\beta_3(y) = 4y$. □

The next example highlights the proposed result on a nonlinear dynamical system.

Example 3.2 Consider the following nonlinear dynamical system:

$$\begin{aligned} \dot{x}_1 &= x_2 x_3 \\ \dot{x}_2 &= x_1 + x_3^2 \\ \dot{x}_3 &= x_2 - x_3 \\ y &= h(x) = x_3. \end{aligned}$$

The observability 1-forms are as follows:

$$\begin{aligned} \theta_1 &= dx_3 \\ \theta_2 &= dx_2 - dx_3 \\ \theta_3 &= dx_1 - dx_2 + (2x_3 + 1)\, dx_3. \end{aligned}$$

Obviously, these 1-forms $\theta = [\theta_1, \theta_2, \theta_3]^T$ are linearly independent, thus the dynamical system satisfies the observability rank condition. Similar to the linear case, we can then calculate the frame $\tau = [\tau_1, \tau_2, \tau_3]$. The solution of the following algebraic equations

$$\begin{aligned} \theta_1 \tau_1 &= 0 \\ \theta_2 \tau_1 &= 0 \\ \theta_3 \tau_1 &= 1 \end{aligned}$$

yields

$$\tau_1 = \frac{\partial}{\partial x_1}.$$

Therefore, by induction we obtain

$$\tau_2 = [\tau_1, f] = \frac{\partial}{\partial x_2}$$

and

$$\tau_3 = [\tau_2, f] = \frac{\partial}{\partial x_3} + x_3 \frac{\partial}{\partial x_1}.$$

The above calculations give the matrix Λ as

$$\Lambda = \theta\tau = \begin{pmatrix} \theta_1\tau_1 & \theta_1\tau_2 & \theta_1\tau_3 \\ \theta_2\tau_1 & \theta_2\tau_2 & \theta_2\tau_3 \\ \theta_3\tau_1 & \theta_3\tau_2 & \theta_3\tau_3 \end{pmatrix} = \begin{pmatrix} 0 & 0 & 1 \\ 0 & 1 & -1 \\ 1 & -1 & 3x_3 + 1 \end{pmatrix},$$

which yields the 1-forms ω as

$$\omega = \begin{pmatrix} \omega_1 \\ \omega_2 \\ \omega_3 \end{pmatrix} = \Lambda^{-1} \begin{pmatrix} \theta_1 \\ \theta_2 \\ \theta_3 \end{pmatrix},$$

where

$$\Lambda^{-1} = \begin{pmatrix} -3x_3 & 1 & 1 \\ 1 & 1 & 0 \\ 1 & 0 & 0 \end{pmatrix}.$$

Thus we have

$$\omega = \begin{pmatrix} 1 & 0 & -x_3 \\ 0 & 1 & 0 \\ 0 & 0 & 1 \end{pmatrix} = \begin{pmatrix} dx_1 - x_3 dx_3 \\ dx_2 \\ dx_3 \end{pmatrix}.$$

By integrating the above equation, the desired change of coordinates is as follows:

$$z = \begin{pmatrix} x_1 - \frac{1}{2}x_3^2 \\ x_2 \\ x_3 \end{pmatrix}$$

via which the studied system can be transformed into the following normal form with output injection:

$$\begin{cases} \dot{z}_1 = 0 \\ \dot{z}_2 = z_1 + \frac{3}{2}y^2 \\ \dot{z}_3 = z_2 + 0 \\ y = z_3. \end{cases}$$

□

3.3 Extension to Systems with Inputs

The result presented in Sect. 3.2 only treated dynamical systems without inputs, while in practice most of systems contain the control inputs. In order to extend this result to systems with inputs, let us consider the following single-input system:

$$\begin{aligned} \dot{x} &= f(x) + g(x)u \\ y &= h(x), \end{aligned} \tag{3.15}$$

where $x \in \mathscr{X} \subseteq \mathbb{R}^n$, $u \in \mathbb{R}$, $y \in \mathbb{R}$, $f : \mathscr{X} \to \mathbb{R}^n$, $g : \mathscr{X} \to \mathbb{R}^n$. It is assumed that f, g and h are smooth, $f(0) = 0$, $g(0) = 0$ and $h(0) = 0$.

Theorem 3.3 *System (3.15) can be transformed, via a local diffeomorphism, into the following normal form with output injection:*

$$\begin{aligned} \dot{z} &= A_O z + \beta(y) + \eta(y)u \\ y &= C_O z, \end{aligned} \tag{3.16}$$

where A_O and C_O are of Brunovsky form, $\beta(y)$ and $\eta(y)$ are functions of y with appropriate dimension, if and only if

(1) one of the conditions in Theorem 3.2 is fulfilled;
(2) for $1 \leq i \leq n-1$, the conditions $[g, \tau_i] = 0$ are satisfied.

Proof Based on the results presented in Theorem 3.2, we can state that there exists a local diffeomorphism $z = \phi(x)$ such that

$$\phi_*(f) = A_O z + \beta(y)$$

and

$$\phi_n(x) = z_n$$

if and only if one of the conditions in Theorem 3.2 is fulfilled. Thus, it remains to prove that the Jacobian of this diffeomorphism $\phi(x)$ will transform $g(x)$ of (3.15) into $\eta(y)$ of (3.16). For this, we need to investigate the value of $\frac{\partial \phi(x)}{\partial x} g := \phi_*(g)$.

Note that the deduced diffeomorphism $\phi(x)$ satisfies $\phi_*(\tau_i) = \omega(\tau_i) = \frac{\partial}{\partial z_i}$ for $1 \leq i \leq n-1$, where ω is defined in (3.11). Then we have

$$\frac{\partial}{\partial z_i}\phi_*(g) = \left[\phi_*(g), \frac{\partial}{\partial z_i}\right] = [\phi_*(g), \phi_*(\tau_i)] = \phi_*([g, \tau_i]),$$

which implies that

$$\frac{\partial}{\partial z_i}\phi_*(g) = 0,$$

if and only if $[g, \tau_i] = 0$ for $1 \leq i \leq n-1$. Note that $\frac{\partial}{\partial z_i}\phi_*(g) = 0$ is equivalent to state that $\phi_*(g)$ is only a function of y, noted as $\eta(y)$. Finally, we proved $g(x)$ in (3.15) can be transformed into $\eta(y)$ in (3.16) if and only if $[g, \tau_i] = 0$ for $1 \leq i \leq n-1$. □

The following simple example illustrates how to apply this result to dynamical systems with single input and single output.

Example 3.3 Consider the following nonlinear dynamical system:

$$\begin{aligned}\dot{x}_1 &= -x_1x_3 - x_2x_3^2 - x_2^2 + (x_3 - x_2)u \\ \dot{x}_2 &= x_1 + x_2x_3 \\ \dot{x}_3 &= x_2 + u \\ y &= x_3.\end{aligned}$$

A straightforward calculation leads to

$$\begin{aligned}\theta_1 &= dx_3 \\ \theta_2 &= dx_2 \\ \theta_3 &= dx_1 + x_3dx_2 + x_2dx_3.\end{aligned}$$

Therefore, we have

$$\begin{aligned}\tau_1 &= \frac{\partial}{\partial x_1} \\ \tau_2 &= \frac{\partial}{\partial x_2} - x_3\frac{\partial}{\partial x_1} \\ \tau_3 &= \frac{\partial}{\partial x_3} - x_2\frac{\partial}{\partial x_1}.\end{aligned}$$

It is easy to verify that

$$[\tau_1, \tau_2] = [\tau_1, \tau_3] = [\tau_2, \tau_3] = 0.$$

Since

$$g = (x_3 - x_2)\frac{\partial}{\partial x_1} + \frac{\partial}{\partial x_3},$$

then we can check that

$$[\tau_1, g] = [\tau_2, g] = 0.$$

Thus, according to Theorem 3.3, there exists a change of coordinates that transforms the above dynamical system into the normal form (3.16). Indeed, a simple calculation yields

$$\begin{aligned}\omega_1 &= dz_1 = d(x_1 + x_2x_3) \\ \omega_2 &= dz_2 = dx_2 \\ \omega_3 &= dz_3 = dx_3.\end{aligned}$$

By integrating the above equation, we can then obtain the following diffeomorphism:

$$z = \begin{bmatrix} z_1 \\ z_2 \\ z_3 \end{bmatrix} = \begin{pmatrix} x_1 + x_2 x_3 \\ x_2 \\ x_3 \end{pmatrix},$$

which transforms the above dynamical system into the following normal form:

$$\begin{aligned} \dot{z}_1 &= yu \\ \dot{z}_2 &= z_1 \\ \dot{z}_3 &= z_2 + u \\ y &= z_3. \end{aligned}$$

□

Theorem 3.3 gave the necessary and sufficient conditions to transform a single-input single-output nonlinear system to an observer normal form with output injection. This result can be easily generalized to treat multiple-input nonlinear systems as well. Consider now the following system:

$$\begin{aligned} \dot{x} &= f(x) + \textstyle\sum_{i=1}^{m} g_i(x)u_i \\ y &= h(x), \end{aligned} \tag{3.17}$$

where $u_i \in \mathbb{R}$, $g_i : \mathscr{X} \to \mathbb{R}^n$ for $1 \leq i \leq m$ is assumed to be smooth with $g_i(0) = 0$. Then, the following corollary can be stated.

Corollary 3.1 *System (3.17) can be transformed, via a local diffeomorphism, into the following normal form with output injection:*

$$\begin{aligned} \dot{z} &= A_O z + \beta(y) + \textstyle\sum_{i=1}^{m} \eta_i(y)u_i \\ y &= C_O z, \end{aligned} \tag{3.18}$$

where A_O and C_O are of Brunovsky form, $\beta(y)$ and $\eta_i(y)$ are functions of y with appropriate dimension, if and only if

1. *one of conditions in Theorem 3.2 is fulfilled;*
2. *for $1 \leq i \leq n-1$ and $1 \leq j \leq m$, the condition $[g_j, \tau_i] = 0$ is satisfied.*

Proof The proof of the above corollary is quite similar to that of Theorem 3.3. □

Remark 3.1 Up to now, we have treated the nonlinear dynamical systems without inputs (Theorem 3.2), with single input (Theorem 3.3) and with multiple inputs (Corollary 3.1). It can be seen that the drift term f and the control term g have been treated separately. More precisely, the desired diffeomorphism $z = \phi(x)$ is mainly determined by the drift term f. Then, additional conditions on the control term g can be added to treat system with single input or multiple inputs. In other words, the control term g has no influence on the deduction of the desired change of coordinates

$z = \phi(x)$. Consequently, in order to focus on the different techniques we are going to apply to deduce the diffeomorphism, the rest of this book will only treat nonlinear dynamical systems without inputs. As we have explained, those results can be then easily adapted for systems with inputs by simply adding complementary conditions on the control terms.

3.4 Observer Design

3.4.1 *Luenberger-Like Observer*

Since the normal forms (3.3), (3.16) and (3.18) have the similar structures, containing both linear terms ($A_O z$ and $C_O z$) and nonlinear terms of known variables (y and u), the designed observers for these three normal forms possess also similar structures. For the sake of generality, let us consider the normal form (3.18) for a general nonlinear dynamical systems with multiple inputs. For this normal form, different types of observers can be designed, such as those presented in Chap. 1. Here, we will design the simplest asymptotic observer: Luenberger-like observer.

Consider the following dynamics:

$$\dot{\hat{z}} = A_O \hat{z} + \beta(y) + \sum_{i=1}^{m} \eta_i(y) u_i + K(y - C_O \hat{z}). \tag{3.19}$$

Note $e = z - \hat{z}$, then according to (3.18) and (3.19) we obtain

$$\dot{e} = (A_O - K C_O) e.$$

Due to the fact that the pair (A_O, C_O) is observable, there exists a K such that $(A_O - KC_O)$ is Hurwitz. To seek such a K, we can place directly the poles of the matrix $(A_O - KC_O)$. Consequently, e is asymptotically stable, and we have

$$\lim_{t \to +\infty} \hat{z}(t) = z(t).$$

Finally, we can obtain the estimation of x by inverting the diffeomorphism $\phi(x)$. In summary, the following dynamics

$$\begin{cases} \dot{\hat{z}} = A_O \hat{z} + \beta(y) + \sum_{i=1}^{m} \eta_i(y) u_i + K(y - C_O \hat{z}) \\ \hat{x} = \phi^{-1}(\hat{z}) \end{cases} \tag{3.20}$$

is an asymptotic Luenberger-like observer of (3.17).

3.4.2 Design Procedure

The following summarizes the procedure to design observer based on the proposed normal form with output injection:

Step 1: According to (3.2), calculate the associated 1-forms θ for the studied system;
Step 2: Determine the family of vector fields τ defined in (3.4) and (3.5);
Step 3: Calculate Λ according to (3.7);
Step 4: Compute ω according to (3.11), then integrate it to obtain the diffeomorphism $z = \phi(x)$;
Step 5: Apply this diffeomorphism to calculate $\phi_*(f)$ and $\phi_*(g)$ where f and g are given by (3.17);
Step 6: Design the proposed Luenberger-like observer (3.20) to estimate x of (3.17).

Exercises

Exercise 3.1 Consider the following dynamical system:

$$\begin{cases} \dot{x}_1 = -x_3^3 \\ \dot{x}_2 = x_1 + 2x_2x_3 \\ \dot{x}_3 = x_2 + x_3^2 \\ y = h(x) = x_3 \end{cases}$$

and note

$$f = -x_3^3 \frac{\partial}{\partial x_1} + (x_1 + 2x_2x_3) \frac{\partial}{\partial x_2} + \left(x_2 + x_3^2\right) \frac{\partial}{\partial x_3}$$

as the vector field that generates the above dynamics.

(1) Compute $L_f h$, $L_f^2 h$ and compare them to $\dot{y}$ and $\ddot{y}$.
(2) Show that $\theta_1 = dh$, $\theta_2 = dL_f h$ and $\theta_3 = dL_f^2 h$ are independent 1-forms.
(3) Compute τ_1 as the solution of the following algebraic equation:

$$\begin{cases} \theta_1(\tau_1) = \theta_2(\tau_1) = 0 \\ \theta_3(\tau_1) = 0. \end{cases}$$

(4) Compute $\tau_2 = [\tau_1, f]$ and $\tau_3 = [\tau_2, f]$.
(5) Show that those vector fields commute, i.e.,

$$[\tau_1, \tau_2] = [\tau_1, \tau_3] = [\tau_2, \tau_3] = 0.$$

What can we deduce from this fact?
(6) Construct the following matrix:

$$\Lambda = \begin{pmatrix} \theta_1(\tau_1) & \theta_1(\tau_2) & \theta_1(\tau_3) \\ \theta_2(\tau_1) & \theta_2(\tau_2) & \theta_2(\tau_3) \\ \theta_3(\tau_1) & \theta_3(\tau_2) & \theta_3(\tau_3) \end{pmatrix}.$$

(7) Compute $dz = \omega = \Lambda^{-1} \begin{pmatrix} \theta_1 \\ \theta_2 \\ \theta_3 \end{pmatrix}$ and deduce the diffeomorphism $z = \phi(x)$.

(8) Check that in the new coordinates we obtain an observer normal form of the following form:

$$\dot{z} = Az + \beta(y)$$
$$y = z_3.$$

Exercise 3.2 Consider the following dynamical system:

$$\begin{cases} \dot{x}_1 = -x_3^3 \\ \dot{x}_2 = x_1 \\ \dot{x}_3 = x_2 \\ y = h(x) = \sin(x_3). \end{cases}$$

Note

$$f = -x_3^3 \frac{\partial}{\partial x_1} + x_1 \frac{\partial}{\partial x_2} + x_2 \frac{\partial}{\partial x_3}$$

as the vector field that generates the above dynamics.

(1) Answer the same questions (1)–(7) in Exercise 3.1.
(2) What can we deduce?

References

1. Boutat, D.: Geometrical conditions for observer error linearization via $\int 0, 1, ..., (n-2)$. In: Proceedings of 7th IFAC Symposium on Nonlinear Control Systems (2007)
2. Boutat, D.: Extended nonlinear observer normal forms for a class of nonlinear dynamical systems. Int. J. Robust Nonlinear Control **25**(3), 461–474 (2015)
3. Crainic, M., Fernandes, R.L.: Integrability of Lie brackets. Ann. Math. 575–620 (2003)
4. Guay, M.: Observer linearization by output-dependent time-scale transformations. IEEE Trans. Autom. Control **47**(10), 1730–1735 (2002)
5. Isidori, A.: Nonlinear Control Systems. Communications and Control Engineering, 3rd edn. Springer, London (1995)
6. Krener, A., Isidori, A.: Linearization by output injection and nonlinear observers. Syst. Control Lett. **3**(1), 47–52 (1983)
7. Krener, A., Respondek, W.: Nonlinear observers with linearizable error dynamics. SIAM J. Control Optim. **23**(2), 197–216 (1985)
8. Luenberger, D.: An introduction to observers. IEEE Trans. Autom. Control **16**(6), 596–602 (1971)

Chapter 4
Observer Normal Form with Output Injection and Output Diffeomorphism

Abstract For certain nonlinear dynamical systems, the geometric conditions provided in Chap. 3 might not be satisfied, because the desired observer normal form was imposed to have a linear output. This chapter aims at relaxing this constraint by proposing new geometric conditions [1]. Those conditions enable us to deduce a change of coordinates that transforms a given nonlinear dynamical system into an observer normal form with nonlinear output [3]. For such a normal form, a high-gain observer will be designed, and the related design procedure will be given at the end of this chapter.

4.1 Problem Statement

Consider the following single-output dynamical system:

$$\begin{aligned} \dot{x} &= f(x) \\ y &= h(x), \end{aligned} \tag{4.1}$$

where $x \in \mathscr{X} \subseteq \mathbb{R}^n$ with $\mathscr{X}$ being a neighborhood of 0, and $y \in \mathbb{R}$. Without loss of generality it is assumed that f and h are smooth with $f(0) = 0$ and $h(0) = 0$.

As we have already discussed in Chap. 3, system (4.1) can be transformed into the observer normal form with output injection (3.3) if and only if the commutativity condition (3.6) of Theorem 3.1 is satisfied. In general, this commutativity condition is restrictive, which can be shown by the following example.

Example 4.1 Consider the following Lotka–Volterra model [2] by assuming per capita growth rates are linear:

$$\begin{aligned} \dot{x}_1 &= x_1(a - bx_2) \\ \dot{x}_2 &= x_2(-c + ex_1) \\ y &= x_1, \end{aligned} \tag{4.2}$$

D. Boutat and G. Zheng, *Observer Design for Nonlinear Dynamical Systems*, Lecture Notes in Control and Information Sciences 487,
https://doi.org/10.1007/978-3-030-73742-9_4

where x_1 and x_2 are the states which are assumed to be positive, y is the output and the parameters a, b, c and e are all positive with the following meanings:

- a is the growth rate of the prey in the absence of interaction with predator;
- b measures the impact of predation;
- c is the natural death of predator in the absence of food;
- e denotes the efficiency of the predator in interaction with prey.

Following Definition 3.1 in Chap. 3, the observability 1-forms for the studied system (4.2) are given by

$$\begin{aligned}\theta_1 &= dy = dx_1\\ \theta_2 &= (a - bx_2)\,dx_1 - bx_1 dx_2,\end{aligned}$$

and it is clear that θ_1 is independent of θ_2. Therefore, the dynamical system fulfills the observability rank condition.

Now, let us follow the procedure presented in Chap. 3 to construct the family of vector field τ. By solving the algebraic equation defined in (3.4), we obtain

$$\tau_1 = -\frac{1}{bx_1}\frac{\partial}{\partial x_2}.$$

Then τ_2 can be computed by the Lie bracket defined in (3.5), and this yields

$$\tau_2 = [\tau_1, f] = \frac{\partial}{\partial x_1} - \frac{a - bx_2 - c + ex_1}{bx_1}\frac{\partial}{\partial x_2},$$

where $f = \begin{bmatrix} x_1(a - bx_2) \\ x_2(-c + ex_1) \end{bmatrix}$.

Then, it is easy to check that

$$[\tau_1, \tau_2] = \frac{2}{x_1}\tau_1 \neq 0,$$

therefore the condition (3.6) of Theorem 3.1 is not satisfied, and we cannot find a change of coordinates $z = \phi(x)$ that transforms dynamical system (4.2) into the nonlinear observer normal form (3.3).

However, if we apply the following change of coordinates

$$\begin{pmatrix} z_1 \\ z_2 \end{pmatrix} = \begin{pmatrix} c\ln x_1 - ex_1 - bx_2 \\ \ln x_1 \end{pmatrix}$$

to system (4.2), a simple calculation yields

$$\begin{aligned}\dot{z}_1 &= \tfrac{d}{dt}(c\ln x_1 - ex_1 - bx_2) = ac - eax_1 = ac - eay\\ \dot{z}_2 &= (a - bx_2) = z_1 - c\ln y + ey\\ \bar{y} &= z_2.\end{aligned}$$

It is worth noting that the new output $\bar{y} = \ln(y)$ is linear in the new coordinates. And the above transformed system can be written into the following nonlinear observer normal form with nonlinear output:

$$\begin{aligned}\dot{z}_1 &= \beta_1(y)\\ \dot{z}_2 &= z_1 + \beta_2(y)\\ y &= \varphi(z_2),\end{aligned}$$

where φ is a diffeomorphism on the output, and

$$\begin{aligned}\beta_1(y) &= ac - eay\\ \beta_2(y) &= -c\ln y + ey\\ \varphi(z_2) &= e^{z_2}.\end{aligned}$$

□

The above example clearly shows that, even if the commutativity condition (3.6) of Theorem 3.1 is not satisfied, it is still possible to transform a nonlinear dynamical system into the normal form (3.3) with nonlinear output, and this is achieved by applying a diffeomorphism $z = \phi(x)$ to the state, which implicitly applies as well a diffeomorphism $\varphi(y)$ to the output.

Therefore, this chapter focuses on how to calculate such a diffeomorphism $\varphi(y)$ on the output, together with a local change of coordinates $z = \phi(x)$, such that the studied dynamical system (4.1) can be transformed into the following nonlinear observer normal form with nonlinear output diffeomorphism:

$$\begin{aligned}\dot{z} &= A_O z + \beta(\bar{y})\\ \bar{y} &= C_O z = z_n,\end{aligned} \tag{4.3}$$

where $\bar{y} = \varphi(y)$, $A_O = \begin{pmatrix} 0 & \cdots & 0 & 0 & 0\\ 1 & \cdots & 0 & 0 & 0\\ \vdots & \ddots & \ddots & \cdots & \vdots\\ 0 & \cdots & 1 & 0 & 0\\ 0 & \cdots & 0 & 1 & 0\end{pmatrix}$ and $C_O = \begin{pmatrix} 0 & \cdots & 0 & 1\end{pmatrix}$.

4.2 Diffeomorphism on Output

For the studied system (4.1), let us follow the same procedure presented in Chap. 3 to calculate firstly the 1-forms

$$\theta = \begin{bmatrix}\theta_1\\ \vdots\\ \theta_n\end{bmatrix}$$

with

$$\theta_i = dL_f^{i-1}h, \text{ for } 1 \leq i \leq n,$$

which is assumed as well to be linearly independent. Therefore, all states of system (4.1) are observable since the observability rank condition is satisfied.

Based on the above 1-forms, in Chap. 3, the family of vector fields τ was determined by (3.4) and (3.5). Then an invertible matrix $\Lambda = \theta\tau$ has been deduced with the special form described in (3.10), i.e.,

$$\Lambda = \theta\tau = \begin{pmatrix} 0 & 0 & \cdots & 0 & \theta_1\tau_n \\ 0 & 0 & \cdots & \theta_2\tau_{n-1} & * \\ \vdots & \vdots & \iddots & \vdots & \vdots \\ 0 & \theta_{n-1}\tau_2 & \cdots & \cdots & * \\ \theta_n\tau_1 & * & \cdots & \cdots & * \end{pmatrix}, \tag{4.4}$$

where

$$\theta_n\tau_1 = \cdots = \theta_1\tau_n = 1.$$

This is due to the construction of τ_1 by fulfilling the algebraic equation (3.4). Then, a simple calculation shows that its inverse can be written as the following form:

$$\Lambda^{-1} = \begin{pmatrix} * & * & \cdots & * & \theta_n\tau_1 \\ * & * & \cdots & \theta_n\tau_1 & 0 \\ \vdots & \vdots & \iddots & \vdots & \vdots \\ * & \theta_n\tau_1 & \cdots & \cdots & 0 \\ \theta_n\tau_1 & 0 & \cdots & \cdots & 0 \end{pmatrix}, \tag{4.5}$$

where $*$ just implies any values which satisfy $\Lambda^{-1}\Lambda = I$. In Theorem 3.1, it has been proven that the differential of the desired diffeomorphism $z = \phi(x)$ was defined as

$$dz = d\phi(x) = \Lambda^{-1}\theta.$$

Therefore, the last component of the diffeomorphism satisfies

$$dz_n = [\theta_n(\tau_1)]\theta_1.$$

Recall that the vector field τ_1 in Chap. 3 was determined by (3.4), satisfying $\theta_n(\tau_1) = 1$ and $\theta_k(\tau_1) = 0$ for $1 \leq k \leq n-1$. Moreover, according to Definition 3.1, we have $\theta_1 = dh$. Consequently, this construction of τ_1 implicitly implies that the deduced diffeomorphism has $z_n = h(x)$. Equivalently, it implies that the transformed system will yield a linear output $y = z_n$.

The above analysis explained why the output of the transformed system is not affected by the deduced diffeomorphism, and more importantly it shows the clue how to introduce a diffeomorphism on the output, i.e., modify the algebraic equation (3.4) which was used to determine τ_1 in Chap. 3.

Following this thought, let us determine the first vector field σ_1 by the following algebraic conditions

$$\begin{cases} \theta_k(\sigma_1) = 0, \text{ for } 1 \le k \le n-1 \\ \theta_n(\sigma_1) = l(y), \end{cases} \tag{4.6}$$

where $l(y)$ is a nowhere-vanishing function of y which will be determined later. It is clear that (4.6) is equivalent to (3.4) when $l(y) = 1$. In this case, we have $\sigma_1 = \tau_1$. For the case $l(y) \neq 1$, σ_1 can be seen as a generalization of τ_1 by relaxing the condition $\theta_n(\tau_1) = 1$, allowing such a nowhere-vanishing function $l(y)$.

Denote $\tau = [\tau_1, \ldots, \tau_n]$ as the family of vector fields determined by (3.4) and (3.5). It is clear that the new constructed vector field σ_1 satisfies

$$\sigma_1 = l(y)\tau_1. \tag{4.7}$$

Similar to (3.5), from the vector field σ_1, we can then define the following family of independent vector fields by induction:

$$\sigma_i = [\sigma_{i-1}, f], \text{ for } 2 \le i \le n. \tag{4.8}$$

As the function $l(y)$ is assumed to be nowhere-vanishing in its domain and τ_i are independent, then the vector fields σ_i are independent as well.

Suppose now the commutativity conditions are not satisfied for τ, i.e., $\left[\tau_i, \tau_j\right] \neq 0$ for $1 \le i, j \le n$, then the key point is to seek such a nowhere-vanishing function $l(y)$ to make the commutativity conditions be satisfied for the vector fields of σ, i.e., $\left[\sigma_i, \sigma_j\right] = 0$ for $1 \le i, j \le n$.

Before stating the differential geometric conditions on the existence of such a diffeomorphism on the output, it is useful to understand the relations between the new family of vector fields σ and the former family of vector fields τ deduced in Chap. 3.

Lemma 4.1 *The vector field τ and the vector field σ are linked by the following equation:*

$$\sigma_k = \sum_{j=0}^{k-1} (-1)^j \binom{k-1}{j} \left(L_f^j l\right) \tau_{k-j}, \tag{4.9}$$

where $\binom{k-1}{j}$ are the binomial coefficients, and $L_f^{j+1} l = L_f L_f^j l$ for $j \ge 0$ with $L_f^0 l = l$.

Proof The proof will be performed by induction. Before that, let us check the first three equations with $k = 1, 2, 3$. For $k = 1$, according to (4.9), it is obvious that

$$\sigma_1 = L_f^0 l(y)\tau_1 = l(y)\tau_1,$$

which is always true by the definition of σ_1 in (4.6).

When $k = 2$, (4.9) reads as

$$\begin{aligned}\sigma_2 &= \sum_{j=0}^{1}(-1)^j\binom{1}{j}\left(L_f^j l\right)\tau_{2-j} \\ &= l(y)\tau_2 - (L_f l)\tau_1.\end{aligned} \tag{4.10}$$

On the other hand, σ_2 is constructed by following (4.8), and we have

$$\begin{aligned}\sigma_2 &= [\sigma_1, f] = [l\tau_1, f] \\ &= L_{l\tau_1} f - L_f(l\tau_1) \\ &= \tfrac{\partial f}{\partial x} l\tau_1 - \tfrac{\partial l}{\partial x}\tau_1 f - l\tfrac{\partial \tau_1}{\partial x} f \\ &= l[\tau_1, f] - (L_f l)\tau_1 \\ &= l\tau_2 - (L_f l)\tau_1.\end{aligned} \tag{4.11}$$

Therefore, from (4.10) and (4.11) we can conclude that (4.9) is satisfied for $k = 2$.

For the case $k = 3$, (4.9) becomes

$$\sigma_3 = l\tau_3 - 2L_f(l)\tau_2 + L_f^2(l)\tau_1.$$

By following (4.8), we have

$$\begin{aligned}\sigma_3 &= [\sigma_2, f] = [l\tau_2 - L_f(l)\tau_1, f] \\ &= [l\tau_2, f] - [L_f(l)\tau_1, f] \\ &= l[\tau_2, f] - L_f\tau_2 + L_f^2\tau_1 - L_f(l)[\tau_1, f] \\ &= l\tau_3 - 2L_f(l)\tau_2 + L_f^2(l)\tau_1.\end{aligned}$$

Thus, the relation (4.9) is true as well for $k = 3$.

For the general case, we are going to proceed by induction. For this, it is assumed that the formula (4.9) is true for $k \geq 1$, and we will show that it is true as well for $k + 1$.

Suppose that σ_k satisfies (4.9), then σ_{k+1} can be calculated by following (4.8) as

$$\begin{aligned}\sigma_{k+1} &= [\sigma_k, f] \\ &= \left[\sum_{j=0}^{k-1}(-1)^j\binom{k-1}{j}\left(L_f^j l\right)\tau_{k-j}, f\right] \\ &= \sum_{j=0}^{k-1}(-1)^j\binom{k-1}{j}\left(L_f^j l\right)[\tau_{k-j}, f] - L_f\left(\sum_{j=0}^{k-1}(-1)^j\binom{k-1}{j}\left(L_f^j l\right)\right)\tau_{k-j}.\end{aligned}$$

Replacing $[\tau_{k-j}, f]$ in the above equation by τ_{k-j+1} and $L_f L_f^j$ by L_f^{j+1}, we obtain

$$\sigma_{k+1} = \sum_{j=0}^{k-1}(-1)^j \binom{k-1}{j} \left(L_f^j l\right) \tau_{k-j+1} - \sum_{j=0}^{k-1}(-1)^j \binom{k-1}{j} \left(L_f^{j+1} l\right) \tau_{k-j}.$$

Rewriting the index in the second sum of the above equation, we obtain

$$\sigma_{k+1} = \sum_{j=1}^{k-1}(-1)^j \left(\tbinom{k-1}{j} + \tbinom{k-1}{j-1}\right) \left(L_f^j l\right) \tau_{k-j} + (-1)^k (L_f^k l)\tau_1 + l\tau_{k+1}. \quad (4.12)$$

Finally, with the following binomial expansions using Pascal's triangle

$$\binom{k-1}{j} + \binom{k-1}{j-1} = \binom{k}{j}.$$

Equation (4.12) can be read as

$$\sigma_{k+1} = \sum_{j=0}^{k}(-1)^j \binom{k}{j} \left(L_f^j l\right) \tau_{k-j},$$

which is exactly Eq. (4.9) for $k+1$, and we proved Lemma 4.1 by induction. □

The relation between σ_i and τ_i has been exploited in (4.9) of Lemma 4.1, which depends on the expression of function $L_f^i l(y)$. Therefore, before computing the Lie bracket $[\sigma_i, \sigma_j]$ for $1 \leq i, j \leq n$, we should firstly know how to compute $L_f^i l(y)$. To do so, we adopt the following result for high-order derivative of the composition of functions given by Faà di Bruno's Formula [7]. To state it, we denote $P_m = \{1, 2, \ldots, m\}$, $P_m^m = P_m \times \cdots \times P_m$ and

$$U_m = \left\{ k = (k_1, \ldots, k_m) \in P_m^m : \sum_{i=1}^{m} i k_i = m \right\}.$$

Based on the above definitions, we denote then by $k! = k_1! \cdots k_m!$ and $|k| = \sum_{i=1}^{m} k_i$. Hence, the successive Lie derivative of $l(y)$ can be achieved by combinatorial calculus, which is stated in the following lemma.

Lemma 4.2 *The mth Lie derivative of the function $l(y) = l(h(x))$ in the direction of the vector field f is given by the following formula:*

$$L_f^m l = \sum_{k \in U_m}^{m} \frac{m!}{k!} l^{(|k|)} \prod_{i=1}^{m} \left(\frac{L_f^i h}{i!} \right)^{k_i}, \quad (4.13)$$

where $l^{(k)} = \frac{d^k l(y)}{dy^k}$ denotes the kth derivative of $l(y)$ with respect to y.

To give an idea how the above formula works, let us calculate the first three high-order derivatives of l in the direction of f

- $L_f l = l'(y) L_f h$
- $L_f^2 l = l''(y)(L_f h)^2 + l'(y) L_f^2 h$
- $L_f^3 l = l^{(3)}(y)(L_f h)^3 + 3l''(y)(L_f h) L_f^2 h + l'(y) L_f^3 h.$

Based on the result stated in Lemma 4.2, we would like to highlight the following facts.

Remark 4.1 If $k_1 = k_2 = \cdots = k_{m-1} = 0$ and $k_m = 1$, i.e., $k = 1$, then it is easy to see from (4.13) that the expression $L_f^m l$ contains the term $l'(y) L_f^m h$. Hence, we have

(1) $L_{\tau_k} L_f^m l = 0$ if $m < n - k$, since $L_{\tau_k} L_f^i h = 0$ for $i \leq m < n - k$;
(2) $L_{\tau_k} L_f^m l = l'(y)$ if $m = n - k$, because $L_{\tau_k} L_f^i h = 0$ for $i < m$.

To check that the commutativity of vector fields σ_i for $1 \leq i \leq n$, it is natural to verify step by step. Thus, we will firstly check the commutativity of σ_1 with the other vector fields, and then pass to σ_2 and so on. Let us start with the vector field σ_1.

Lemma 4.3 *The following items are equivalent:*

(1) $[\sigma_1, \sigma_k] = 0,$ *for* $2 \leq k \leq n - 1;$
(2) $[\tau_1, \tau_k] = 0,$ *for* $2 \leq k \leq n - 1.$

Moreover, we have

$$[\sigma_1, \sigma_n] = l^2 [\tau_1, \tau_n] + \left((-1)^{n-1} - 1\right) l \frac{dl}{dy} \tau_1. \tag{4.14}$$

Proof Using the result stated in Lemma 4.1, we have

$$\sigma_k = \sum_{i=0}^{k-1} (-1)^i \binom{k-1}{i} (L_f^i l) \tau_{k-i}, \text{ for } 2 \leq k \leq n.$$

From Remark 4.1, we know that $L_{\tau_1} L_f^i l = 0$ for $i < n$ and $L_{\tau_k} l = 0$ for $k < n$. Therefore, for $k < n$ we have

$$[\sigma_1, \sigma_k] = \sum_{i=0}^{k-1} (-1)^i \binom{k-1}{i} l (L_f^i l) [\tau_1, \tau_{k-i}].$$

The following will use the above results to prove Lemma 4.3 by induction. For this, consider firstly $k = 2$.

- For the case $n = 2$ and $k = 2$, then from the expression of σ_2 in Lemma 4.1, we have

$$\begin{aligned} \sigma_1 &= l \tau_1 \\ \sigma_2 &= l \tau_2 - (L_f l) \tau_1. \end{aligned}$$

A straightforward calculation gives

$$\begin{aligned}[\sigma_1, \sigma_2] &= \left[l\tau_1, l\tau_2 - (L_f l)\tau_1\right] \\ &= l^2\,[\tau_1, \tau_2] - ll'dh(\tau_2)\tau_1 - ll'dL_f h\,(\tau_1)\,\tau_1 \\ &= l^2\,[\tau_1, \tau_2] - 2ll'\tau_1\end{aligned}$$

because $dh(\tau_2) = 1$ and $dL_f h\,(\tau_1) = 1$, which is exactly the formula (4.14) for $n = 2$;

- For the case $n > 2$ and $k = 2$, we obtain

$$[\sigma_1, \sigma_2] = l^2[\tau_1, \tau_2].$$

Therefore, as $l(y) \neq 0$, we can state that $[\sigma_1, \sigma_2] = 0$ if and only if $[\tau_1, \tau_2] = 0$.

With the last equation obtained above, we can then step forward to consider the case $k = 3$. By following the same way, it is easy to show that $[\sigma_1, \sigma_3] = 0$ if and only if $[\tau_1, \tau_3] = 0$. This process is iterated up to $k = n - 1$.

Now, assume that $[\tau_1, \tau_k] = 0$ for $k \leq n - 1$, then

$$[\sigma_1, \sigma_n] = l^2[\tau_1, \tau_n] - L_{\tau_n}(l)\tau_1 + (-1)^{n-1}L_{\tau_1}(L_f^{n-1}l)\tau_1,$$

which yields, thanks to Remark 4.1, the following equation:

$$[\sigma_1, \sigma_n] = l^2[\tau_1, \tau_n] + \left((-1)^{n-1} - 1\right)ll'\tau_1,$$

and it is exactly formula (4.14). □

The following corollary is a direct result of Lemma 4.3.

Corollary 4.1 *Assume that the equivalent items* (1) *and* (2) *of Lemma 4.3 are fulfilled.*

(1) If n is an even integer, i.e., $n = 2p$ for some $p \in \mathbb{N}$, then

$$[\sigma_1, \sigma_n] = 0,$$

if and only if

$$[\tau_1, \tau_n] = \mu_1(y)\tau_1.$$

In this case, we obtain

$$l(y) = e^{\frac{1}{2}\int_0^y \mu_1(s)ds},$$

which is one solution of the differential equation $l^2\mu_1 - 2ll' = 0$, deduced from (4.14);

(2) If n is an odd integer, i.e., $n = 2p + 1$ for some $p \in \mathbb{N}$, then

$$[\sigma_1, \sigma_n] = 0,$$

if and only if

$$[\tau_1, \tau_n] = 0.$$

Example 4.2 Let us consider again Example 4.1 to show how to compute the function l. From the previous calculations, we know that

$$[\tau_1, \tau_2] = \frac{1}{x_1}\tau_1,$$

thus in the above differential equation $\mu = \frac{2}{x_1} = \frac{2}{y}$. Therefore, according to Corollary 4.1, we get

$$l(y) = e^{\int_0^y \frac{ds}{ys}} = y.$$

Now set $\sigma_1 = l\tau_1$, then

$$\sigma_2 = [\sigma_1, f] = l\tau_2 - L_f l\tau_1.$$

Therefore,

$$[\sigma_1, \sigma_2] = [l\tau_1, l\tau_2 - L_f l\tau_1] = l^2[\tau_1, \tau_2] - lL_{\tau_2}l\tau_1 - lL_{\tau_1}(L_f l)\tau_1.$$

As $l = y = x_1$, then

$$L_f l = x_1(a - bx_2),$$

which implies that $L_{\tau_1}(L_f l) = 1$. We also have $L_{\tau_2}l = 1$ and $[\tau_1, \tau_2] = \frac{2}{l}$, then we deduce

$$[\sigma_1, \sigma_2] = 0.$$

□

We saw that, if n is an odd integer, i.e., $n = 2p + 1$ for some $p \in \mathbb{N}$, then Eq. (4.14) is reduced to $[\sigma_1, \sigma_n] = l^2[\tau_1, \tau_n]$. Therefore, $[\sigma_1, \sigma_n] = 0$ if and only if $[\tau_1, \tau_n] = 0$. In this case, since the Lie bracket $[\sigma_1, \sigma_n]$ does not give us information about $l(y)$, we should study other Lie brackets, such as $[\sigma_2, \sigma_n]$, to deduce $l(y)$. Inspired by this thought, the following lemma is to investigate the commutativity of σ_2 with σ_n for the case where n is odd.

Lemma 4.4 *Assume $n = 2p + 1$ and $[\tau_1, \tau_k] = 0$ for all $2 \le k \le n$, then the following items are equivalent:*

(1) $[\sigma_2, \sigma_k] = 0$ *for all* $2 \le k \le n - 1$*;*
(2) $[\tau_2, \tau_k] = 0$ *for all* $2 \le k \le n - 1$.

Moreover, we have

$$[\sigma_2, \sigma_n] = l^2[\tau_2, \tau_n] - nl\frac{dl}{dy}\tau_2 + \gamma(x)\tau_1. \tag{4.15}$$

Proof Recall that the expressions of σ_2 and σ_3 are

$$\sigma_2 = l\tau_2 - L_f(l)\tau_1$$

and

$$\sigma_3 = l\tau_3 - 2L_f(l)\tau_2 + L_f^2(l)\tau_1.$$

If $n = 3$, then from the above two expressions we obtain

$$[\sigma_2, \sigma_3] = l^2[\tau_2, \tau_3] + \left(-2lL_{\tau_2}L_f(l) - lL_{\tau_3}l\right)\tau_2 + \gamma\tau_1,$$

where

$$\gamma = -3l'L_f(l) + lL_{\tau_2}L_f^2(l) + lL_{\tau_3}L_f(l).$$

A straightforward calculation leads to

$$[\sigma_2, \sigma_3] = l^2[\tau_2, \tau_3] - 3ll'\tau_2 + \gamma\tau_1.$$

Now, we assume that $n = 2p + 1 > 3$, then we have the following cases:

- If $2 \leq k \leq n-2$, then $[\sigma_2, \sigma_k] = 0$ if and only if $[\tau_2, \tau_k] = 0$. Indeed, recall from above that

$$\sigma_k = \sum_{j=0}^{k-1}(-1)^j\binom{k-1}{j}\left(L_f^j l\right)\tau_{k-j}.$$

  As $\sigma_2 = l\tau_2 - L_f(l)\tau_1$, then we have

$$[\sigma_2, \sigma_k] = [l\tau_2, \sigma_k] - \left[L_f(l)\tau_1, \sigma_k\right].$$

  We also know that, for $0 \leq j \leq k \leq n-2$,

$$L_{\tau_2}L_f^j l = 0, \quad L_{\tau_1}L_f^j l = 0, \quad L_{\tau_k}l = 0$$

  and

$$L_{\tau_k}L_f l = 0.$$

  As $[\tau_1, \tau_k] = 0$, we obtain

$$[\sigma_2, \sigma_k] = l\,[\tau_2, \sigma_k] = \sum_{j=0}^{k-1}(-1)^j\binom{k-1}{j}\left(L_f^j l\right)\left[\tau_2, \tau_{k-j}\right].$$

  Therefore, for $k = 3 \leq n-2$ we have

$$[\sigma_2, \sigma_k] = l\,[\tau_2, \sigma_k] = l\,[\tau_2, \tau_3] = 0.$$

Finally, we can prove the result by induction as follows: If $[\sigma_2, \sigma_s] = [\tau_2, \tau_s] = 0$ for all $s \leq k-1$, then a straightforward calculation leads to $\left[\sigma_2, \sigma_{s+1}\right] = l\left[\tau_2, \tau_{s+1}\right]$. Thereby, $[\tau_2, \tau_k] = 0$ for all $k \leq n-2$ if and only if $[\sigma_2, \sigma_k] = 0$ for all $k \leq n-2$.

- If $k = n-1$, then we obtain

$$\begin{aligned}\left[\sigma_2, \sigma_{n-1}\right] &= l^2\left[\tau_2, \tau_{n-1}\right] + (-1)^{n-2} L_{\tau_2}\left(L_f^{n-2} l\right)\tau_1 + L_{\sigma_{n-1}}\left(L_f l\right)\tau_1 \\ &= l^2\left[\tau_2, \tau_{n-1}\right] + \left((-1)^{n-2} l l' + l l'\right)\tau_1.\end{aligned}$$

Thus, we have

$$[\sigma_2, \sigma_{n-1}] = l^2[\tau_2, \tau_{n-1}] + l l'(-1 + (-1)^{n-1})\tau_1.$$

As n is odd, this leads to $[\sigma_2, \sigma_{n-1}] = 0$ if and only if $[\tau_2, \tau_{n-1}] = 0$.

Therefore, for $2 \leq k \leq n-1$, we proved that the item (1) is equivalent to the item (2) of Lemma 4.4.

Finally, consider σ_k defined in (4.9). Taking $k = n$, by keeping only the non-zero terms, we have

$$\begin{aligned}[\sigma_2, \sigma_n] = l\,[\tau_2, \tau_n] + \left[l\tau_2, (-1)^{n-2}\binom{n-1}{n-2}\left(L_f^{n-2} l\right)\tau_2 + (-1)^{n-1}\left(L_f^{n-1} l\right)\tau_1\right] \\ -l\left(L_{\tau_n} l\right)\tau_2.\end{aligned}$$

Therefore, after some simplifications we obtain

$$[\sigma_2, \sigma_n] = l\,[\tau_2, \tau_n] - n l l'\tau_2 + \gamma\tau_1,$$

where $\gamma = -n l'\left(L_f l\right) + l L_{\tau_n}\left(L_f l\right) + l L_{\tau_2}\left(L_f^{n-1} l\right)$. □

Like the case of even dimension (i.e., $n = 2p$) where the function $l(y)$ can be obtained via $[\sigma_1, \sigma_n] = 0$, the similar result can be stated for the case of odd dimension.

Corollary 4.2 *Assume that the conditions stated in Lemma 4.4 are fulfilled, then* $[\sigma_2, \sigma_n] = 0$ *only if* $[\tau_2, \tau_n] = \mu_2(y)\tau_2 + \mu_1(x)\tau_1$. *In this case, we obtain*

$$l(y) = e^{\frac{1}{n}\int_0^y \mu_2(s)ds},$$

which is one solution of $l^2\mu_2(y) - n l l' = 0$.

Remark 4.2 Corollary 4.2 provides only a necessary condition to ensure $[\sigma_2, \sigma_n] = 0$. Precisely, substituting this necessary condition $[\tau_2, \tau_n] = \mu_2(y)\tau_2 + \mu_1(x)\tau_1$ back into (4.15), we have

$$\begin{aligned}[\sigma_2, \sigma_n] &= l^2[\tau_2, \tau_n] - n l\frac{dl}{dy}\tau_2 + \gamma(x)\tau_1 \\ &= \left(l^2\mu_2(y) - n l l'\right)\tau_2 + l^2\mu_1(x)\tau_1.\end{aligned}$$

It is clear that, if $l(y)$ is chosen as the solution of $l^2\mu_2(y) - nll' = 0$, we have

$$[\sigma_2, \sigma_n] = l^2\mu_1(x)\tau_1,$$

which shows that such a choice of $l(y)$ can only ensure $[\sigma_2, \sigma_n] = 0$ when $\mu_1(y) = 0$. This issue will be highlighted in the following example.

Example 4.3 Consider the following dynamical system:

$$\begin{aligned}\dot{x}_1 &= x_3x_1\\ \dot{x}_2 &= x_1 - 3x_3\\ \dot{x}_3 &= x_2\\ y &= x_3.\end{aligned}$$

A straightforward calculation provides

$$\begin{aligned}\theta_1 &= dx_3\\ \theta_2 &= dx_2\\ \theta_3 &= dx_1 - 3dx_3.\end{aligned}$$

Therefore, the vector fields of the frame τ are given by

$$\begin{aligned}\tau_1 &= \tfrac{\partial}{\partial x_1}\\ \tau_2 &= \tfrac{\partial}{\partial x_2} + x_3\tfrac{\partial}{\partial x_1}\\ \tau_3 &= \tfrac{\partial}{\partial x_3} + x_3\tfrac{\partial}{\partial x_2} + \left(x_3^2 - x_2\right)\tfrac{\partial}{\partial x_1}.\end{aligned}$$

These vector fields fulfill the following equations:

$$\begin{aligned}[\tau_1, \tau_2] &= 0\\ [\tau_1, \tau_3] &= 0\\ [\tau_2, \tau_3] &= -\tau_1.\end{aligned}$$

Comparing to the necessary condition stated in Corollary 4.2, i.e.,

$$[\tau_2, \tau_3] = \mu_2(y)\tau_2 + \mu_1(x)\tau_1,$$

we have

$$\begin{aligned}\mu_2(y) &= 0\\ \mu_1(x) &= -1.\end{aligned}$$

Solving the equation $l^2\mu_2(y) - nll' = 0$, we obtain

$$l(y) = d_{\text{const}}.$$

Thus we have

$$[\sigma_2, \sigma_n] = l^2\mu_1(x)\tau_1 = -d^2_{\text{const}}\tau_1 = -d^2_{\text{const}}\frac{\partial}{\partial x_1} \neq 0.$$

This example shows that the stated problem cannot be solved by means of a diffeomorphism on the output. We will see later in Chap. 5 that this kind of systems can be transformed into another observer normal form, thanks to the so-called immersion technique. □

4.3 Diffeomorphism for States

It is clear now that the information of the existence of such a function $l(y)$ is hidden in two Lie brackets, either $[\tau_1, \tau_n]$ or $[\tau_2, \tau_n]$, depending on the dimension (odd or even) of the studied system. In the case where the state dimension n is even, then the information of $l(y)$ is hidden in $[\tau_1, \tau_n]$. For n is odd, then $l(y)$ is contained in $[\tau_2, \tau_n]$. These two Lie brackets enable us to determine the nowhere-vanishing function $l(y)$.

After having determined the function $l(y)$, we can then calculate the family of vector fields $\sigma = [\sigma_1, \ldots, \sigma_n]$ via (4.7) and (4.8). With the 1-forms $\theta = [\theta_1^T, \ldots, \theta_n^T]^T$ where $\theta_i = dL_f^{i-1}h$ for $1 \leq i \leq n$, we can then calculate the following matrix:

$$\bar{\Lambda} = \theta\sigma = \begin{pmatrix} 0 & 0 & \cdots & 0 & l(y) \\ 0 & 0 & \cdots & l(y) & \bar{\Lambda}_{2,n} \\ \vdots & \vdots & \iddots & \vdots & \vdots \\ 0 & l(y) & \cdots & \bar{\Lambda}_{n-1,n-1} & \bar{\Lambda}_{n-1,n} \\ l(y) & \bar{\Lambda}_{n,2} & \cdots & \bar{\Lambda}_{n,n-1} & \bar{\Lambda}_{n,n} \end{pmatrix}. \tag{4.16}$$

It is clear that this matrix is invertible for nowhere-vanishing function $l(y)$, thus we can define the following 1-forms:

$$\bar{\omega} = \bar{\Lambda}^{-1}\theta := (\bar{\omega}_i)_{1\leq i\leq n}\,. \tag{4.17}$$

Note that the determination of function $l(y)$ enables us to compute uniquely the frame σ, which can then be used to calculate the diffeomorphism $\varphi(y)$ on the output, according to the results stated in Corollaries 4.1 and 4.2. Based on the matrix $\bar{\Lambda}$ defined in (4.16), we can then state the result on the deduction of the diffeomorphism $z = \phi(x)$ in a similar way as Theorem 3.2.

Theorem 4.1 *The following three assertions are equivalent:*

(1) *There exists a local change of coordinates $z = \phi(x)$ that transforms dynamical system (4.1) into the nonlinear observer normal form (4.3);*
(2) *The vector fields σ_i and σ_j for $1 \leq i, j \leq n$ defined in (4.7)–(4.8) commute, i.e.,*

$$\left[\sigma_i, \sigma_j\right] = 0; \tag{4.18}$$

(3) The 1-forms $\bar{\omega}_i$ defined in (4.17) for $1 \le i \le n$ are closed, i.e.,

$$d\bar{\omega}_i = 0. \tag{4.19}$$

Furthermore, the change of coordinates is determined by

$$z_i = \phi_i(x) = \int_\gamma \bar{\omega}_i, \tag{4.20}$$

where $\gamma : [0, 1] \to \mathscr{X}$ is any smooth curve on contractile neighborhood of 0 such that $\gamma(0) = 0$ and $\gamma(1) = x$.

Proof The proof is similar to that of Theorem 3.2 in Chap. 3 by replacing τ_i by σ_i. The only difference is the diffeomorphism on the output. Therefore, the following will focus only on the deduction of this diffeomorphism $\varphi(y)$.

For the matrix $\bar{\Lambda}$ defined in (4.16), a simple calculation shows that its inverse is of the following form:

$$\bar{\Lambda}^{-1} = \begin{pmatrix} * & * & \cdots & * & \frac{1}{l(y)} \\ * & * & \cdots & \frac{1}{l(y)} & 0 \\ \vdots & \vdots & \iddots & \vdots & \vdots \\ * & \frac{1}{l(y)} & \cdots & \cdots & 0 \\ \frac{1}{l(y)} & 0 & \cdots & \cdots & 0 \end{pmatrix}. \tag{4.21}$$

Since

$$dz = d\phi(x) = \bar{\Lambda}^{-1}\theta,$$

thus the last component of the diffeomorphism satisfies

$$dz_n = \frac{1}{l(y)}\theta_1 = \frac{1}{l(y)}dy,$$

where we use the definition of θ_1 with $\theta_1 = dh = dy$. Finally, we obtain the diffeomorphism on the output as

$$z_n = \varphi(y) = \int_0^y \frac{ds}{l(s)}. \tag{4.22}$$

□

Theorem 4.1 shows the constructive way to calculate the diffeomorphism $z = \phi(x)$ whose last component is of the form $z_n = \varphi(y) = \int_0^y \frac{ds}{l(s)}$. Precisely, we first build the matrix $\bar{\Lambda}$ defined in (4.16) by evaluating the 1-forms θ_i on the vector fields σ_i. Then, we compute the components of the change of coordinates as the integral of the components of $\bar{\Lambda}^{-1}\bar{\omega}$ where the components of $\bar{\omega}$ are given in (4.17). The following example is to highlight the above result.

Example 4.4 Consider the following system:

$$\begin{aligned}\dot{x}_1 &= \alpha_{1,1}(x_2)x_1 + \alpha_{1,2}(x_2)x_1^2 + \beta_1(x_2)\\ \dot{x}_2 &= \alpha_{2,2}(x_2)x_1 + \beta_2(x_2)\\ y &= x_2,\end{aligned} \tag{4.23}$$

where $\alpha_{2,2}(x_2) \neq 0$. A straightforward calculation gives us the observability 1-forms as follows:

$$\begin{aligned}\theta_1 &= dx_2\\ \theta_2 &= \alpha_{2,2}(x_2)dx_1 + \left(\alpha_{2,2}'(x_2)x_1 + \beta_2'(x_2)\right)dx_2.\end{aligned}$$

Thereby, we deduce the frame τ as

$$\begin{aligned}\tau_1 &= \frac{1}{\alpha_{2,2}}\frac{\partial}{\partial x_1}\\ \tau_2 &= \frac{\partial}{\partial x_2} + \frac{1}{\alpha_{2,2}}\left(\frac{\alpha_{2,2}'}{\alpha_{2,2}}(\alpha_{2,2}x_1 + \beta_2) + \alpha_{1,1}(x_2) + 2\alpha_{1,2}(x_2)x_1\right)\frac{\partial}{\partial x_1}.\end{aligned}$$

It can be checked that these two vector fields do not commute. Indeed, we have

$$[\tau_1, \tau_2] = \left(2\frac{\alpha_{2,2}'}{\alpha_{2,2}^2} + 2\frac{\alpha_{1,2}}{\alpha_{2,2}^2}\right)\frac{\partial}{\partial x_1} = 2\left(\frac{\alpha_{2,2}'}{\alpha_{2,2}} + \frac{\alpha_{1,2}}{\alpha_{2,2}}\right)\tau_1.$$

According to Corollary 4.1, we obtain

$$l(y) = e^{\int_0^y \frac{\alpha_{2,2}'(s)}{\alpha_{2,2}(s)}ds + \int_0^y \frac{\alpha_{1,2}(s)}{\alpha_{2,2}^2(s)}ds} = \alpha_{2,2}(y)e^{\int_0^y \frac{\alpha_{1,2}(s)ds}{\alpha_{2,2}(s)}}.$$

Hence

$$\sigma_1 = l(y)\tau_1 = e^{\int_0^y \frac{\alpha_{1,2}(s)ds}{\alpha_{2,2}(s)}}\frac{\partial}{\partial x_1},$$

and we obtain

$$\begin{aligned}\sigma_2 &= [\sigma_1, f]\\ &= e^{\int_0^y \frac{\alpha_{1,2}(s)ds}{\alpha_{2,2}(s)}}\left(\alpha_{2,2}\frac{\partial}{\partial x_2} + \left(\alpha_{1,1} - \frac{\alpha_{1,2}}{\alpha_{2,2}}\left(\alpha_{2,2}x_1 + \beta_2\right) + 2\alpha_{1,2}x_1\right)\frac{\partial}{\partial x_1}\right).\end{aligned}$$

Then, it is easy to check that

$$[\sigma_1, \sigma_2] = 0.$$

This last condition ensures the existence of the diffeomorphism $z = \phi(x)$. To construct it, we first build $\bar{\Lambda}$ as follows:

$$\bar{\Lambda} = \begin{pmatrix}\theta_1(\sigma_1) & \theta_1(\sigma_2)\\ \theta_2(\sigma_1) & \theta_2(\sigma_2)\end{pmatrix} = \begin{pmatrix}0 & l(y)\\ l(y) & \bar{\Lambda}_{22}\end{pmatrix},$$

where $l(y) = \alpha_{2,2}(y)e^{\int_0^y \frac{\alpha_{1,2}(s)ds}{\alpha_{2,2}(s)}}$ and

$$\bar{\Lambda}_{22} = l\left(\alpha_{1,1} - \frac{\alpha_{1,2}}{\alpha_{2,2}}\left(\alpha_{2,2}x_1 + \beta_2\right) + 2\alpha_{1,2}x_1\right) + l\left(\alpha_{2,2}'x_1 + \beta_2'\right).$$

Since the inverse of $\bar{\Lambda}$ is

$$\bar{\Lambda}^{-1} = \begin{pmatrix} -\frac{\bar{\Lambda}_{22}}{l^2} & \frac{1}{l} \\ \frac{1}{l} & 0 \end{pmatrix},$$

then we obtain

$$\omega_2 = \tfrac{1}{l}\theta_1 = \tfrac{dx_2}{l(x_2)}$$

and

$$\begin{aligned} \omega_1 &= -\tfrac{\bar{\Lambda}_{22}}{l^2}\theta_1 + \tfrac{1}{l}\theta_2 \\ &= -\tfrac{1}{l}\left(\left(\alpha_{1,1} - \tfrac{\alpha_{1,2}}{\alpha_{2,2}}\beta_2 + \alpha_{1,2}x_1\right)\right)dx_2 + \tfrac{1}{l}\alpha_{2,2}dx_1 \\ &= d\left(\tfrac{1}{l}\alpha_{2,2}x_1 + \int_0^y \varsigma(s)ds\right), \end{aligned}$$

where

$$\varsigma = -\frac{1}{l}\left(\alpha_{1,1} - \frac{\alpha_{1,2}}{\alpha_{2,2}}\beta_2\right).$$

Finally, we obtain the following diffeomorphism:

$$z = \phi(x) = \begin{bmatrix} \frac{1}{l}\alpha_{2,2}x_1 + \int_0^y \varsigma(s)ds \\ \int_0^y \frac{ds}{l(s)} \end{bmatrix}$$

via which the studied system (4.23) can be transformed into the sought normal form. □

4.4 Observer Design

4.4.1 High-Gain Observer

By considering the normal form (4.3), similar to the normal form (3.18) in Chap. 3, different types of observers can be designed. Here, we will design a high-gain asymptotic observer [4, 5] for such a normal form.

Consider the following dynamics:

$$\dot{\hat{z}} = A_O\hat{z} + \beta(\bar{y}) + \Delta_\varepsilon K(\varphi(y) - C_O\hat{z}), \tag{4.24}$$

where $\bar{y} = \varphi(y)$ and

$$\Delta_\varepsilon = \begin{bmatrix} \varepsilon^n & 0 & \cdots & 0 \\ 0 & \varepsilon^{n-1} & \cdots & 0 \\ \vdots & \vdots & \ddots & \vdots \\ 0 & 0 & \cdots & \varepsilon \end{bmatrix}$$

with ε being large positive constant and $K = [K_1, \dots, K_n]^T$ with K_i being chosen such that $(A_O - KC_O)$ is Hurwitz, which is always possible since the pair (A_O, C_O) is observable.

Note $e_i = z_i - \hat{z}_i$, then according to (4.3) and (4.24), we obtain

$$\dot{e}_i = e_{i-1} - \varepsilon^{n-i+1} K_i e_n. \tag{4.25}$$

Denote $\bar{e} = [\bar{e}_1, \dots, \bar{e}_n]^T$ and

$$\bar{e}_i = \frac{1}{\varepsilon^{n-i}} e_i$$

for $1 \leq i \leq n$, then we have

$$\begin{aligned} \dot{\bar{e}}_i &= \tfrac{1}{\varepsilon^{n-i}} \dot{e}_i = \tfrac{1}{\varepsilon^{n-i}} \left[e_{i-1} - \varepsilon^{n-i+1} K_i e_n \right] \\ &= \tfrac{1}{\varepsilon^{n-i}} e_{i-1} - \varepsilon K_i e_n \\ &= \varepsilon \bar{e}_{i-1} - \varepsilon K_i \bar{e}_n \\ &= \varepsilon \left[\bar{e}_{i-1} - K_i \bar{e}_n \right] \end{aligned} \tag{4.26}$$

which is equivalent to

$$\dot{\bar{e}} = \varepsilon \left[A_O - K C_O \right] \bar{e}.$$

Due to the fact that $A_O - KC_O$ is Hurwitz and ε is positive, $\bar{e}$ is asymptotically stable, so is e. Consequently, we have

$$\lim_{t \to +\infty} \hat{z}(t) = z(t).$$

Therefore, we can obtain the estimation of x by inverting the diffeomorphism $\phi(x)$. In summary, the following dynamics

$$\begin{aligned} \dot{\hat{z}} &= A_O \hat{z} + \beta(\bar{y}) + \Delta_\varepsilon K (\varphi(y) - C_O \hat{z}) \\ \hat{x} &= \phi^{-1}(\hat{z}) \end{aligned} \tag{4.27}$$

is an asymptotic observer of (4.1).

Compared to Luenberger-like observer designed in (3.20), an additional diagonal matrix Δ_ε is adopted when designing observer (4.27). The main reason is to attenuate the influence of noise in the model. In such a situation, the observation error dynamics of (4.25) can be rewritten as

$$\dot{e}_i = e_{i-1} - \varepsilon^{n-i+1} K_i e_n + d_i,$$

where d_i represents the noise (or uncertainties which cannot be modeled). Following the same procedure, we can then obtain

$$\begin{aligned} \dot{\bar{e}}_i &= \frac{1}{\varepsilon^{n-i}} \dot{e}_i = \frac{1}{\varepsilon^{n-i}} \left[e_{i-1} - \varepsilon^{n-i+1} K_i e_n + d_i \right] \\ &= \varepsilon \left[\bar{e}_{i-1} - K_i \bar{e}_n \right] + \frac{1}{\varepsilon^{n-i}} d_i \end{aligned} \tag{4.28}$$

which is equivalent to

$$\dot{\bar{e}} = \varepsilon \left[A_O - K C_O \right] \bar{e} + \bar{\Delta}_\varepsilon d,$$

where $d = [d_1, \ldots, d_n]^T$ and

$$\bar{\Delta}_\varepsilon = \begin{bmatrix} \frac{1}{\varepsilon^{n-1}} & 0 & \cdots & 0 \\ 0 & \frac{1}{\varepsilon^{n-2}} & \cdots & 0 \\ \vdots & \vdots & \ddots & \vdots \\ 0 & 0 & \cdots & 1 \end{bmatrix}.$$

Due to the fact that ε is chosen to be large positive constant, the influence of the noise d_i is attenuated by multiplying $\frac{1}{\varepsilon^{n-i}}$, defined in $\bar{\Delta}_\varepsilon$ [6].

4.4.2 Design Procedure

The following summarizes the procedure to design observer based on the proposed normal form (4.3):

Step 1: According to (3.2), calculate the associated 1-forms θ for the studied system;

Step 2: Determine the family of vector fields τ defined in (3.4) and (3.5);

Step 3: Calculate the new family of vector fields σ according to (4.7) and (4.8) with unknown function $l(y)$ to be determined;

Step 4: Apply Corollaries 4.1 and 4.2 to compute $l(y)$;

Step 5: Calculate $\bar{\Lambda}$ according to (4.16) and ω according to (4.17), then compute the diffeomorphism $\phi(x)$ according to (4.20);

Step 6: Apply this diffeomorphism to calculate $\phi_*(f)$ where f is given by (4.1);

Step 7: Design the proposed high-gain observer (4.27) to estimate x of (4.1).

Exercises

Exercise 4.1 Consider the dynamical system

$$\begin{cases} \dot{x}_1 = x_1 x_2 \\ \dot{x}_2 = x_1 \\ y = h(x) = \sin x_2, \text{ with } x_2 \in \left]0, \frac{\pi}{2}\right[\end{cases}$$

and note

$$f = x_1 x_2 \frac{\partial}{\partial x_1} + x_1 \frac{\partial}{\partial x_2}$$

as the vector field that generates the above dynamics.

(1) Compute $L_f h$.
(2) Compute $\tau_2 = [\tau_1, f]$.
(3) Show that $[\tau_1, \tau_2] = \varphi(y)\frac{\partial}{\partial x_1} \neq 0$. What can we deduce from this fact?
(4) Set $\sigma_1 = \alpha(y)\tau_1$. Compute $\sigma_2 = [\sigma_1, f]$ and deduce $\alpha(y)$ such that $[\sigma_1, \sigma_2] = 0$.
(5) Construct the matrix

$$\Lambda = \begin{pmatrix} \theta_1(\sigma_1) & \theta_1(\sigma_2) \\ \theta_2(\sigma_1) & \theta_2(\sigma_2) \end{pmatrix}.$$

(6) Compute $dz = \omega = \Lambda^{-1}\begin{pmatrix} \theta_1 \\ \theta_2 \end{pmatrix}$ and deduce the diffeomorphism $z = \phi(x)$.
(7) Check that in this new coordinate we obtain the following observer normal form:

$$\begin{aligned} \dot{z} &= Az + \beta(y) \\ y &= z_2. \end{aligned} \tag{4.29}$$

Exercise 4.2 Consider the following dynamical system:

$$\begin{cases} \dot{x}_1 = -x_3^3 \\ \dot{x}_2 = x_1 \\ \dot{x}_3 = x_2 \\ y = h(x) = \sin(x_3). \end{cases}$$

Use the same procedure as in the above exercise to find an observer normal form similar to (4.29).

Exercise 4.3 Consider the following dynamical system:

$$\begin{cases} \dot{x}_1 = x_1 x_3 \\ \dot{x}_2 = x_1 \\ \dot{x}_3 = x_1 \\ y = h(x) = x_3. \end{cases}$$

Show that it is not possible to find the diffeomorphism for the above dynamical system by using the previous algorithm to transform it into the observer normal form (4.29).

References

1. Boutat, D.: Extended nonlinear observer normal forms for a class of nonlinear dynamical systems. Int. J. Robust Nonlinear Control **25**(3), 461–474 (2015)
2. Boutat, D., Saif, M.: Observer normal forms for a class of Predator-Prey models. J. Frankl. Inst. **353**(10), 2178–2198 (2016)
3. Boutat, D., Benali, A., Hammouri, H., Busawon, K.: New algorithm for observer error linearization with a diffeomorphism on the outputs. Automatica **45**(10), 2187–2193 (2009)
4. Busawon, K., Farza, M., Hammouri, H.: A simple observer for a class of nonlinear systems. Appl. Math. Lett. **11**(3), 27–31 (1998)
5. Gauthier, J.P., Kupka, I.: Deterministic Observation Theory and Applications. Cambridge University Press, Cambridge (2001)
6. Hammouri, H., de Leon Morales, J.: Observer synthesis for state-affine systems. In: Proceedings of the 29th IEEE Conference on Decision and Control, pp. 784–785 (1990)
7. Johnson, W.: The curious history of Faà di Bruno's formula. Am. Math. Mon. **109**(3), 217–234 (2002)

Chapter 5
Observer Normal Form by Means of Extended Dynamics

Abstract This chapter provides a new geometric algorithm to overcome some obstructions that cannot be handled by those results proposed in Chaps. 3 and 4. This result is based on those presented in Chaps. 3 and 4, by adding an auxiliary dynamics into the studied nonlinear systems. Necessary and sufficient geometric conditions are deduced to guarantee the existence of such an auxiliary dynamics and a change of coordinates to transform the nonlinear dynamical system together with that added auxiliary dynamics into an observer normal form [9]. Unlike the normal form (3.3) and (4.3), the desired observer normal form in this chapter will be a function of the output and the auxiliary variables which are used to define the auxiliary dynamics. Moreover, this algorithm enables us to compute the auxiliary dynamics as well as the change of coordinates. This chapter only deals with the case of scalar extended dynamic, and gives an example to illustrate the possibility to do as well high-dimensional extension.

5.1 Problem Statement

Consider the following single-output dynamical system

$$\begin{aligned} \dot{x} &= f(x) \\ y &= h(x), \end{aligned} \tag{5.1}$$

where $x \in \mathscr{X} \subseteq \mathbb{R}^n$ with $\mathscr{X}$ being a neighborhood of 0, and $y \in \mathbb{R}$. Without loss of generality, it is assumed that f and h are smooth with $f(0) = 0$ and $h(0) = 0$.

We have shown in Chap. 3 that such a nonlinear system might be transformed, via a diffeomorphism on the state, into the observer normal form with output injection (3.3) if and only if τ, defined in (3.5), satisfies the commutativity condition (3.6). If not, then we may introduce a diffeomorphism on the output, and transform system (5.1) into the observer normal form (4.3). If unfortunately this solution does not work as well, we can try to introduce an extended dynamics into the studied system to solve such a problem. The following example highlights this idea.

D. Boutat and G. Zheng, *Observer Design for Nonlinear Dynamical Systems*,
Lecture Notes in Control and Information Sciences 487,
https://doi.org/10.1007/978-3-030-73742-9_5

Example 5.1 Consider the following dynamical system

$$\begin{aligned}\dot{x}_1 &= \gamma(y)x_1\\ \dot{x}_2 &= x_1\\ \dot{x}_3 &= x_2\\ y &= x_3\end{aligned} \tag{5.2}$$

with $\gamma'(y) = \frac{d\gamma(y)}{dy} \neq 0$, i.e., $\gamma(y)$ is not constant. A straightforward calculation leads to

$$\begin{aligned}\tau_1 &= \tfrac{\partial}{\partial x_1}\\ \tau_2 &= \tfrac{\partial}{\partial x_2} + \gamma(y)\tau_1\\ \tau_3 &= \tfrac{\partial}{\partial x_3} + \gamma(y)\tau_2 - x_2\gamma'(y)\tau_1.\end{aligned}$$

It is easy to see that

$$[\tau_2, \tau_3] = -2\gamma'(y)\tau_1 \neq 0,$$

therefore Theorem 3.2 in Chap. 3 cannot be applied to transform the studied system into the normal form (3.3) via a local diffeomorphism. Alternatively, as proposed in Chap. 4, we can try to seek firstly a diffeomorphism on the output, and then construct another diffeomorphism on the state to transform the studied system into (4.3). However, since $n = 3$ and $\gamma'(y) \neq 0$, therefore

$$[\tau_2, \tau_3] = 0\tau_2 - 2\gamma'(y)\tau_1.$$

By using Corollary 4.2, we have

$$\begin{aligned}\mu_1(x) &= -2\gamma'(y) \neq 0\\ \mu_2(y) &= 0.\end{aligned}$$

In order to satisfy the equation $l^2\mu_2(y) - nll' = 0$, while $l(y)$ should be a constant not equal to 0. According to Remark 4.2, we have

$$[\sigma_2, \sigma_n] = l^2\mu_1(x)\tau_1 \neq 0,$$

and we can draw the conclusion that Theorem 4.1 cannot be applied as well to transform (5.2) into (4.3).

Nevertheless, if we increase the system's dimension by adding an auxiliary dynamics as follows:

$$\begin{aligned}\dot{x}_1 &= \gamma(y)x_1\\ \dot{x}_2 &= x_1\\ \dot{x}_3 &= x_2\\ \dot{w} &= \kappa(w)\gamma(y)\\ y &= x_3,\end{aligned} \tag{5.3}$$

where $w \in \mathbb{R}$ is an auxiliary variable and $\kappa(w) \neq 0$ which can be chosen freely such that the dynamics of w is bounded. Then it can be seen that, with the following change of coordinates

$$\begin{pmatrix} z_1 \\ z_2 \\ z_3 \\ \xi \end{pmatrix} = \begin{pmatrix} e^{-\int_0^w \frac{ds}{\kappa(s)}} x_1 \\ e^{-\int_0^w \frac{ds}{\kappa(s)}} x_2 + e^{-\int_0^w \frac{ds}{\kappa(s)}} \int_0^y \gamma(s) ds \\ e^{-\int_0^w \frac{ds}{\kappa(s)}} x_3 \\ w \end{pmatrix}$$

system (5.3) can be transformed into the following observer normal form:

$$\begin{aligned} \dot{z}_1 &= 0 \\ \dot{z}_2 &= z_1 - \gamma(y) e^{-\int_0^\xi \frac{ds}{\kappa(s)}} \int_0^y \gamma(s) ds \\ \dot{z}_3 &= z_2 - e^{-\int_0^\xi \frac{ds}{\kappa(s)}} \left(\int_0^y \gamma(s) ds - \gamma(y) y \right) \\ \dot{\xi} &= \kappa(\xi) \gamma(y) \\ \bar{y} &= z_3, \end{aligned} \tag{5.4}$$

where the former output y is linked to the new output by $\bar{y} = e^{-\int_0^\xi \frac{ds}{\kappa(s)}} y$. □

In summary, if the commutativity condition (4.18) of Theorem 4.1 in Chap. 4 cannot be fulfilled, then we might construct an auxiliary dynamics $\dot{w} = \eta(w, y)$ with $w \in \mathbb{R}$ such that the dynamical system (5.1), together with this auxiliary dynamics, admits a diffeomorphism $(z^T, \xi)^T = \phi(x, w)$ which transforms the extended dynamical system

$$\begin{aligned} \dot{x} &= f(x) \\ \dot{w} &= \eta(w, y) \\ y &= h(x) \end{aligned} \tag{5.5}$$

into the following nonlinear observer form:

$$\begin{aligned} \dot{z} &= A_O z + \beta(\bar{y}, w) \\ \dot{\xi} &= \mu(\xi, \bar{y}) \\ \bar{y} &= z_n = C_O z, \end{aligned} \tag{5.6}$$

where the new output $\bar{y}$ is linked to the former one y via a diffeomorphism $\varphi(y, w)$, i.e., $\bar{y} = \varphi(y, w)$. Sometimes, this solution is named as immersion technique in the literature, and the rest of this chapter aims at deducing such a diffeomorphism to transform (5.5) to (5.6).

5.2 Observer Normal Form with Scalar Extended Dynamics

As usual, it is assumed that system (5.1) fulfills the observability rank condition, i.e., the 1-forms $\theta_i = dL_f^{i-1}h$ for $1 \leq i \leq n$ is linearly independent. Denote $\tau = [\tau_1, \ldots, \tau_n]$ as the frame determined by (3.4)–(3.5).

In order to introduce the extended dynamics, denote $w \in \mathbb{R}$ as an auxiliary variable and assume that its dynamics is governed by

$$\dot{w} = \eta(w, y),$$

which will be determined hereafter. Note that $w \in \mathbb{R}$ can be considered as a known variable since it contains only measurable variable y, and the initial condition of w can be chosen freely by us. With this extended dynamics, system (5.1) can be rewritten as

$$\begin{pmatrix} \dot{x} \\ \dot{w} \end{pmatrix} = F(x, w) = \begin{pmatrix} f(x) \\ \eta(w, y) \end{pmatrix}.$$
$$y = h(x)$$

Note F in the following form of vector field

$$F = f + \eta \frac{\partial}{\partial w} \tag{5.7}$$

and let $l(w) \neq 0$ be a nowhere-vanishing function of w which will be determined hereafter. Following the same procedure presented in Chap. 4, we define

$$\sigma_1 = l(w)\tau_1. \tag{5.8}$$

Thereby, we can construct the following family of vector fields by induction:

$$\sigma_i = [\sigma_{i-1}, F]. \tag{5.9}$$

Recall that a result on the relationship between the frame σ and τ has been stated in Lemma 4.1 of Chap. 4 for the dynamical system without extension. The following result is just an adaptation for the extended system.

Lemma 5.1 *The vector field τ and the vector field σ are linked by the following equation:*

$$\sigma_k = \sum_{j=0}^{k-1} (-1)^j \binom{k-1}{j} \left(L_F^j l\right) \tau_{k-j}. \tag{5.10}$$

Proof The proof of this lemma is quite similar to that of Lemma 4.1 in Chap. 4, thus it is left to the readers. □

The main difference between the above result and the one stated in Lemma 4.1 in Chap. 4 is the successive derivatives of $L_F^k l$, where the vector field F is defined in (5.7). Moreover, since $l(w)$ in (5.10) is a function of the auxiliary variable w, therefore it is necessary to understand how to calculate the high-order derivatives of the composition of those functions given in Lemma 5.1.

Proposition 5.1 *For F and τ_i defined in (5.7) and (3.5), the following equations hold*

(1) $L_{\tau_i} L_F^k l = 0$ *for* $1 \leq i \leq n$ *and* $0 \leq k \leq n - i$;
(2) $L_{\tau_i} L_F^{n-i+1} l = l'(w) \frac{\partial \eta}{\partial y}$ *for* $1 \leq i \leq n$ *where* $l'(w) = \frac{dl}{dw}$.

Proof According to the definition of F since

$$F = f + \eta(y, w) \frac{\partial}{\partial w},$$

then we have

$$L_F l = l'(w) \eta(y, w).$$

As the above function depends on the output $y = h(x)$, therefore

$$L_F^2 l = \eta L_{\frac{\partial}{\partial w}} (L_F l) + l'(w) \frac{\partial \eta}{\partial y} L_f h.$$

By induction, for $1 \leq k \leq n - i$, it can be shown that $L_F^k l$ contains a term of the form $l'(w) \frac{\partial \eta}{\partial y} L_f^{k-1} h$. Due to the definition of τ_i in (3.5), we have

$$L_{\tau_i} L_f^k h = 0 \text{ for } 0 \leq k \leq n - i$$

and

$$L_{\tau_i} L_f^{n-i} h = 1.$$

Using the above two equations, we can then obtain the expected result, and prove that $L_{\tau_i} L_F^k l = 0$ and $L_{\tau_i} L_F^{n-i+1} l = l'(w) \frac{\partial \eta}{\partial y}$ for $1 \leq i \leq n$ and $0 \leq k \leq n - i$. □

Based on Proposition 5.1, the following will show how to deduce a differential equation to determine the function $l(w)$. Before this, let us make the following assumption.

Assumption 5.1 For system (5.1) with dimension n,

(1) if n is even, i.e., $n = 2p$ for some $p \in \mathbb{N}$, then it is assumed that $[\tau_1, \tau_k] = 0$ for all $2 \leq k \leq n$;
(2) if n is odd, i.e., $n = 2p + 1$ for some $p \in \mathbb{N}$, then it is assumed that $[\tau_1, \tau_k] = 0$ for all $1 \leq k \leq n$, $[\tau_2, \tau_k] = 0$ for all $2 \leq k \leq n - 1$ and $[\tau_2, \tau_n] = \mu_1(y)\tau_1$ where $\mu_1(y)$ is a function of the output y.

The above assumption just means that we have already tried those two methods presented in Chaps. 3 and 4 which allow us to reach the stage where the above assumption is satisfied. Similar to the approach used in Chap. 4, we can then determine the nowhere-vanishing function $l(w)$ by calculating the Lie brackets between σ_i and σ_j.

Theorem 5.1 *Under Assumption 5.1, the following assertions hold.*

(1) If $n = 2p + 1$ *and* $[\tau_2, \tau_n] = \mu_1(y)\tau_1$, *then* $[\sigma_2, \sigma_n] = 0$ *if and only if there exist* $l(w)$ *and* $\eta(y, w)$ *that fulfill the following differential equation:*

$$l\mu_1 + 2l'\eta_y = 0, \tag{5.11}$$

where $\eta_y = \frac{\partial\eta}{\partial y}$;

(2) If $n = 2p$ *and* $[\tau_3, \tau_n] = \mu_2(y)\tau_2 + \mu_1\tau_1$, *then* $[\sigma_3, \sigma_n] = \nu\sigma_1$ *if there exists* $l(w)$ *that fulfills the following differential equation:*

$$l\mu_2 + (n+1)l'\eta_y = 0, \tag{5.12}$$

where $\eta_y = \frac{\partial\eta}{\partial y}$.

Proof Consider firstly the case $n = 2p + 1$. We know from Lemma 5.1 that

$$\sigma_2 = l\tau_2 - L_F l\tau_1,$$

thus we have

$$\begin{aligned}[\sigma_2, \sigma_n] &= \left[\sigma_2, \sum_{i=0}^{n-1}(-1)^i \binom{n-1}{j}\left(L_F^i l\right)\tau_{n-i}\right] \\ &= l\left(L_{\tau_n}(L_F l) + (-1)^{n-1} L_{\tau_2}\left(L_F^{n-1} l\right)\right)\tau_1 + l^2[\tau_2, \tau_n],\end{aligned}$$

which, after simplification, yields

$$[\sigma_2, \sigma_n] = l^2[\tau_2, \tau_n] + l\left(l'\eta_y + (-1)^{n-1} l'\eta_y\right)\tau_1,$$

where $\eta_y = \frac{\partial\eta}{\partial y}$.

Now, if $[\tau_2, \tau_n] = \mu_1(y)\tau_1$, then $[\sigma_2, \sigma_n] = 0$ if and only if there exist $\eta(y, w)$ and $l(w)$ such that the following equation is fulfilled

$$l\mu_1 + 2l'\eta_y = 0.$$

For the case $n = 2p$, since $[\sigma_2, \sigma_n] = 0$ if and only if $[\tau_2, \tau_n] = 0$, therefore, we need to calculate the Lie bracket $[\sigma_3, \sigma_n]$ to get the information about $l(w)$. From Lemma 5.1, we obtained

$$\sigma_3 = l\tau_3 - 2L_F l\tau_2 + L_F^2 l\tau_1,$$

thus its Lie bracket with σ_n reads as

$$\begin{aligned}[\sigma_3, \sigma_n] &= \left[\sigma_3, \sum_{i=0}^{n-1} (-1)^i \binom{n-1}{j} \left(L_F^i l\right) \tau_{n-i}\right] \\ &= l^2 [\tau_3, \tau_n] + \left(2l L_{\tau_n} (L_F l) + l (-1)^{n-2} \binom{n-1}{n-2} L_{\tau_3} \left(L_F^{n-2} l\right)\right) \tau_2 + \mu\tau_1.\end{aligned}$$

A straightforward calculation leads to

$$[\sigma_3, \sigma_n] = l^2 [\tau_3, \tau_n] + \left(2ll'\eta_y + (n-1)\, ll'\eta_y\right) \tau_2 + \mu\tau_1,$$

which can be simplified as follows:

$$[\sigma_3, \sigma_n] = l^2 [\tau_3, \tau_n] + (n+1)\, ll'\eta_y\tau_2 + \mu\tau_1.$$

In order to eliminate the term containing τ_2 in the above equation, if we assume that

$$[\tau_3, \tau_n] = \mu_2(y)\tau_2 + \mu_1\tau_1,$$

then it leads to the following equation:

$$l\mu_2(y) + (n+1)l'\eta_y = 0,$$

and it ends the proof of Theorem 5.1. □

Remark 5.1 Theorem 5.1 states that the expected solutions of $l(w)$ and $\eta(y, w)$ should satisfy either (5.11) or (5.12), depending on the dimension of the studied system. It is evident that those solutions might not be unique. In general,

(1) for the case $n = 2p + 1$, we might take the solution of (5.11) as

$$\eta(y, w) = \kappa(w)\psi(y) + \kappa_1(w),$$

where $\kappa(w) \neq 0$, $\frac{d\psi(y)}{dy} = \mu_1(y)$ and

$$l(w) = e^{-\frac{1}{2}\int_0^w \frac{ds}{\kappa(s)}};$$

(2) for the case $n = 2p$, we might take the solution of (5.12) as

$$\eta(y, w) = \kappa(w)\psi(y) + \kappa_1(w),$$

where $\kappa(w) \neq 0$, $\frac{d\psi(y)}{dy} = \mu_2(y)$ and

$$l(w) = e^{-\frac{1}{n+1}\int_0^w \frac{ds}{\kappa(s)}}.$$

Remark 5.2 As we have seen in the proof of Theorem 5.1 for the case $n = 2p$ that the auxiliary dynamics is introduced to eliminate the term containing τ_2 when calculating the Lie bracket $[\sigma_3, \sigma_n]$. Then, we obtain $[\sigma_3, \sigma_n] = \mu_1\sigma_1$. If $\mu_1 = 0$, we need to check other Lie brackets, such as $[\sigma_4, \sigma_n]$.

Remark 5.3 As we did in Chap. 4, after having determined the nowhere-vanishing function $l(w)$, we can then compute the new vector field σ_1 via (5.8). Following (5.9), we obtain σ_i for $2 \le i \le n$. If their Lie brackets commute, i.e.,

$$\left[\sigma_i, \sigma_j\right] = 0,$$

for $1 \le i, j \le n$, then there always exists another vector field, noted as σ_{n+1}, such that the following properties hold:

(1)
$$\sigma_{n+1} = \frac{\partial}{\partial w} + T,$$

where $T \in \text{span}\{\sigma_1, \ldots, \sigma_n\}$. This implies that σ_{n+1} does not depend on other σ_i for $1 \le i \le n$;

(2) σ_{n+1} commutes with other σ_i, i.e.,

$$[\sigma_i, \sigma_{n+1}] = 0,$$

for $1 \le i \le n$.

The next step is then to evaluate the 1-forms $\theta_i = dL_f^{i-1}h$ for $1 \le i \le n$, together with $\theta_{n+1} = dw$, on the vector fields σ_i for $1 \le i \le n+1$. For this, note

$$\theta = \begin{bmatrix} \theta_1 \\ \vdots \\ \theta_n \\ dw \end{bmatrix}$$

and

$$\sigma = \left[\sigma_1, \ldots, \sigma_n, \sigma_{n+1}\right].$$

Then, we can calculate the following matrix

$$\bar{A} = \theta\sigma = \begin{pmatrix} 0 & 0 & \ldots & 0 & 1 & \bar{A}_{1,n+1} \\ 0 & \vdots & \ldots & 1 & \bar{A}_{2,n} & \bar{A}_{2,n+1} \\ \vdots & \ldots & & \ldots\ldots & \vdots & \vdots \\ 0 & 1 & \ldots & \ldots & \vdots & \vdots \\ 1 & \bar{A}_{n,2} & \ldots & \ldots & \bar{A}_{n,n} & \bar{A}_{n,n+1} \\ 0 & 0 & \ldots & \ldots & 0 & 1 \end{pmatrix} \tag{5.13}$$

with $\bar{\Lambda}_{i,j} = \theta_i\sigma_j$ for $1 \leq i, j \leq n+1$. It is clear that this matrix is invertible, since its $n+1$ columns are linearly independent. Thus, we can define the 1-forms $\bar{\omega}_i$ for $1 \leq i \leq n+1$, which are components of the following 1-forms:

$$\bar{\omega} = \begin{bmatrix} \bar{\omega}_1 \\ \vdots \\ \bar{\omega}_{n+1} \end{bmatrix} = \bar{\Lambda}^{-1}\theta. \tag{5.14}$$

Similar to Theorem 4.1, we can state the following result for the dynamical system with extended dynamics.

Theorem 5.2 *The following three assertions are equivalent:*

(1) There exists a local change of coordinates $(z^T, \xi)^T = \phi(x, w)$ that transforms dynamical system (5.5) into the nonlinear observer normal form (5.6);
(2) The vector field σ_i and σ_j for $1 \leq i, j \leq n+1$ defined in (5.8)–(5.9) commute, i.e.,

$$\left[\sigma_i, \sigma_j\right] = 0; \tag{5.15}$$

(3) The 1-forms $\bar{\omega}_i$ defined in (5.14) for $1 \leq i \leq n+1$ are closed, i.e.,

$$d\bar{\omega}_i = 0. \tag{5.16}$$

Furthermore, the change of coordinates is determined by

$$\begin{aligned} z_i &= \phi_i(x, w) = \textstyle\int_\gamma \bar{\omega}_i \\ \xi &= \phi_{n+1}(x, w) = \textstyle\int_\gamma \bar{\omega}_{n+1}, \end{aligned} \tag{5.17}$$

where $\gamma : [0, 1] \to \mathscr{X}$ is any smooth curve on contractile neighborhood of 0 such that $\gamma(0) = 0$ and $\gamma(1) = x$.

Proof The proof is similar to that of Theorem 3.2 in Chap. 3 and that of Theorem 4.1 in Chap. 4. □

Example 5.2 (*Example* 5.1 *continued*) For system (5.2), we have obtained

$$\begin{aligned} \tau_1 &= \tfrac{\partial}{\partial x_1} \\ \tau_2 &= \tfrac{\partial}{\partial x_2} + \gamma(y)\tau_1 \\ \tau_3 &= \tfrac{\partial}{\partial x_3} + \gamma(y)\tau_2 - x_2\gamma'(y)\tau_1, \end{aligned}$$

which yields, since $\gamma'(y) \neq 0$, the following equation:

$$[\tau_2, \tau_3] = -2\gamma'(y)\tau_1 \neq 0.$$

Therefore, we have concluded that neither Theorem 3.2 nor Theorem 4.1 can be applied to transform (5.2) into the presented form (3.3) or (4.3).

The following shows how to transform (5.2) into (5.6) by means of the extended dynamics. To do so, let us consider system (5.2) with the extended dynamics as follows:

$$\begin{aligned} \dot{x}_1 &= \gamma(y)x_1 \\ \dot{x}_2 &= x_1 \\ \dot{x}_3 &= x_2 \\ \dot{w} &= \kappa(w)\psi(y) + \kappa_1(w), \end{aligned} \tag{5.18}$$

where $w \in \mathbb{R}$ is an auxiliary and known variable. Since $n = 3$, according to Theorem 5.1, as $[\tau_2, \tau_3] = -2\gamma'(y)\tau_1$, thus

$$\mu_1(y) = -2\gamma'(y).$$

Following the discussions in Remark 5.1, we can choose

$$\eta(y, w) = \kappa(w)\psi(y) + \kappa_1(w),$$

where $\kappa(w) \neq 0$ and

$$\psi(y) = -2\gamma(y),$$

which gives

$$\eta_y = -2\kappa(w)\gamma'(y).$$

Then, (5.11) can be read as

$$l\mu_1 + 2l'\eta_y = -2l\gamma'(y) - 4\kappa(w)l'\gamma'(y) = 0.$$

Due to the fact that $\gamma'(y) \neq 0$, then one solution of the above equation is

$$l(w) = e^{-\frac{1}{2}\int_0^w \frac{ds}{\kappa(s)}}.$$

With the determined nowhere-vanishing function $l(w)$, according to (5.8), we can then define the first vector field σ_1 as follows:

$$\sigma_1 = l\tau_1 = l\frac{\partial}{\partial x_1}.$$

Following (5.9), a straightforward calculation gives

$$\sigma_2 = l\tau_2 - l\gamma\tau_1 = l\frac{\partial}{\partial x_2}$$

and

$$\begin{aligned} \sigma_3 &= l\tau_3 - 2l\gamma\tau_2 + l\left(\gamma' x_2 + \gamma^2\right)\tau_1 \\ &= l\tfrac{\partial}{\partial x_3} - l\gamma\tfrac{\partial}{\partial x_2}. \end{aligned}$$

It can be then checked that

$$[\sigma_2, \sigma_3] = 0.$$

Since the dimension of the extended system is $n + 1 = 4$, we need to construct the 4th vector field that commutes with the other three vector fields σ_1, σ_2 and σ_3. Following the discussion in Remark 5.3, we can see that the following vector field

$$\sigma_4 = \frac{\partial}{\partial w} + \frac{l'}{l} x_1 \frac{\partial}{\partial x_1} + \frac{l'}{l} (x_2 - \gamma x_3 + \psi) \frac{\partial}{\partial x_2} + \frac{l'}{l} x_3 \frac{\partial}{\partial x_3}$$

with

$$\frac{d\psi(y)}{dy} = \gamma(y)$$

satisfies this requirement.

Recall that the observability 1-forms for (5.2) is $\theta_1 = dx_3, \theta_2 = dx_2$ and $\theta_3 = dx_1$. Thus, we have

$$\theta = \begin{bmatrix} \theta_1 \\ \theta_2 \\ \theta_3 \\ dw \end{bmatrix} = \begin{bmatrix} dx_3 \\ dx_2 \\ dx_1 \\ dw \end{bmatrix}$$

and

$$\sigma = [\sigma_1, \sigma_2, \sigma_3, \sigma_4] = \begin{bmatrix} l \; 0 \;\; 0, & \frac{l'}{l} x_1 \\ 0 \; l \; -l\gamma, & \frac{l'}{l} (x_2 - \gamma x_3 + \psi) \\ 0 \; 0 \;\; l, & \frac{l'}{l} x_3 \\ 0 \; 0 \;\; 0, & 1 \end{bmatrix}.$$

Next, we can compute $\bar{\Lambda} = \theta(\sigma)$ as follows:

$$\bar{\Lambda} = \begin{pmatrix} 0 \; 0 \;\; l, & \frac{l'}{l} x_3 \\ 0 \; l \; -l\gamma, & \frac{l'}{l} x_2 - \frac{l'}{l} (\gamma x_3) + \frac{l'}{l} \psi \\ l \; 0 \;\; 0, & \frac{l'}{l} x_1 \\ 0 \; 0 \;\; 0, & 1 \end{pmatrix},$$

and we have

$$\begin{pmatrix} dz_1 \\ dz_2 \\ dz_3 \\ d\xi \end{pmatrix} = \bar{\Lambda}^{-1} \theta = \begin{pmatrix} 0 & 0 & \frac{1}{l}, & -\frac{1}{l^2} l' x_1 \\ \frac{1}{l}\gamma & \frac{1}{l} & 0, & \frac{1}{l^2} \left(-\psi l' - l' x_2 \right) \\ \frac{1}{l} & 0 & 0, & -\frac{1}{l^2} l' x_3 \\ 0 & 0 & 0, & 1 \end{pmatrix} \begin{pmatrix} \theta_1 \\ \theta_2 \\ \theta_3 \\ dw \end{pmatrix}.$$

This leads to the following change of coordinates:

$$
\begin{aligned}
z_1 &= \tfrac{x_1}{l} = e^{-\int_0^w \frac{ds}{\kappa(s)}} x_1 \\
z_2 &= \tfrac{x_2}{l} + \tfrac{\psi}{l} = e^{-\int_0^w \frac{ds}{\kappa(s)}} x_2 + e^{-\int_0^w \frac{ds}{\kappa(s)}} \int_0^y \gamma(s)ds \\
z_3 &= \tfrac{x_3}{l} = e^{-\int_0^w \frac{ds}{\kappa(s)}} x_3 \\
\xi &= w,
\end{aligned}
$$

which transforms the dynamical system (5.2) into the form (5.4). □

5.3 High-Dimensional Extension

Section 5.2 presented how to introduce a scalar auxiliary dynamics to relax the commutativity condition of Lie brackets. If such a scalar extension fails to relax the commutativity condition, then we might try to add high-dimensional auxiliary dynamics. The following presents only an example of two-dimensional extension, and a more general setting for such a method can be found in [1, 2, 4] and references therein.

Consider the following dynamical system

$$
\begin{aligned}
\dot{x}_1 &= \gamma(y)x_2, \quad \dot{x}_2 = x_1 \\
\dot{x}_3 &= x_1, \quad \dot{x}_4 = x_3 \\
y &= x_4.
\end{aligned}
\tag{5.19}
$$

A straightforward calculation leads to

$$
\begin{aligned}
\theta_1 &= dx_4, \quad \theta_2 = dx_3 \\
\theta_3 &= dx_2, \quad \theta_4 = dx_1
\end{aligned}
$$

and

$$
\begin{aligned}
\tau_1 &= \frac{\partial}{\partial x_1}, \quad \tau_2 = \frac{\partial}{\partial x_2} \\
\tau_3 &= \frac{\partial}{\partial x_3} + \gamma \frac{\partial}{\partial x_1} \\
\tau_4 &= \frac{\partial}{\partial x_4} - x_3\gamma' \frac{\partial}{\partial x_1}.
\end{aligned}
$$

Thereby, we obtain

$$
[\tau_1, \tau_2] = [\tau_1, \tau_3] = [\tau_1, \tau_4] = [\tau_2, \tau_3] = 0 \tag{5.20}
$$

and

$$
[\tau_3, \tau_4] = -2\gamma'\tau_1.
$$

We would like to remark that a simple scalar dynamic extension for (5.19) cannot solve the problem to transform the studied system into the proposed nonlinear observer normal form (5.6), and this is due to (5.20) where the Lie brackets are 0 which prevent us from computing $l(w)$. Therefore, we need to seek high-dimensional extension.

For this, let us consider now the following auxiliary dynamics

$$\begin{aligned}\dot{w}_1 &= w_2\\ \dot{w}_2 &= -\gamma.\end{aligned}$$

It can be checked that the following change of coordinates

$$\begin{aligned}z_1 &= e^{-w_1}\left(x_1 + w_2x_2 + w_2^2x_3 + w_2^3x_4 + 2w_2\int_0^y \gamma(s)ds\right)\\ z_2 &= e^{-w_1}\left(x_2 + 3w_2^2x_4 + 2w_2x_3 + 2\int_0^y \gamma(s)ds\right)\\ z_3 &= e^{-w_1}(x_3 + 3w_2x_4)\\ z_4 &= e^{-w_1}x_4\end{aligned}$$

brings the studied example into the following normal form:

$$\begin{aligned}\dot{z}_1 &= -e^{-w_1}\left(w_2^4y + 2w_2^2\int_0^y \gamma(s)ds + 3w_2^2\gamma y + 2\gamma\int_0^y \gamma(s)ds\right)\\ \dot{z}_2 &= z_1 - e^{-w_1}\left(4w_2^3y + 4w_2\int_0^y \gamma(s)ds + 6\gamma w_2 y\right)\\ \dot{z}_3 &= z_2 - e^{-w_1}\left(6w_2^2y + 3\gamma y + 2\int_0^y \gamma(s)ds\right)\\ \dot{z}_4 &= z_3 - 4e^{-w_1}w_2y\\ \dot{w}_1 &= w_2, \quad \dot{w}_2 = -\gamma\\ \bar{y} &= z_4 = e^{-w_1}y.\end{aligned}$$

5.4 Observer Design

5.4.1 Adaptive Observer

The idea of adaptive observer was initialized in [5] to treat the problem of simultaneous state and parameter estimations [8]. In [7], a Kalman-like adaptive observer was presented for state-affine systems with linear time-varying matrix and a high-gain observer was proposed in [6] for the nonlinear system with triangular form. After that, other adaptive observers are studied to estimate of the states for more general normal forms, including the output-depending normal form [3, 10].

Consider the normal form (5.6), this section will design such an adaptive observer for this normal form. Let us consider the following dynamics:

$$\dot{\hat{z}} = A_O\hat{z} + \beta(\bar{y}, w) + K_t(\varphi(y) - C_O\hat{z}), \tag{5.21}$$

where $\bar{y} = \varphi(y)$ and $w = \xi$ are regarded as known measurements, and K_t satisfies the following equation:

$$\begin{aligned} \dot{S} &= -\rho S - SA_O - A_O^T S + C_O^T C_O \\ K_t &= S^{-1}C_O^T, \end{aligned} \tag{5.22}$$

which $\rho > 0$.

Unlike the Luenberger-like and high-gain observers, (5.21) uses a time-varying gain K_t. This observer works as well even the matrices A_O and C_O are time-varying, under the condition that the pair (A_O, C_O) is observable. In this case, (5.21) is similar to the so-called Kalman–Bucy filter without noises in the model and the output. When A_O and C_O are constant matrices of the Brunovsky form, it has been proven that the matrix S defined in (5.22) is symmetric positive definite, and S_∞ (the value of matrix S when $t \to \infty$) is the unique solution of the following Riccati equation

$$0 = \rho S_\infty + S_\infty A_O + A_O^T S_\infty - C_O^T C_O$$

whose components have the similar structure of Δ_ε of the high-gain observer (4.27).

To show the convergence of the proposed observer, let us note $e = z - \hat{z}$. Then according to (5.6) and (5.21), we obtain

$$\dot{e} = A_O e - K_t C_O e$$

with K_t being determined by (5.22). To prove the convergence of e to 0, we can choose the following Lyapunov function:

$$V = e^T Se,$$

then its time derivative equals

$$\begin{aligned} \dot{V} &= \dot{e}^T Se + e^T S\dot{e} + e^T \dot{S}e \\ &= e^T \left[-\rho S - C^T C\right] e \\ &= -\rho V - ||C_O e||_2^2. \end{aligned}$$

Finally, we can state that there exists a positive ρ such that

$$\dot{V} \leq -\rho V,$$

and this implies that e will exponentially converge to 0, i.e.,

$$\lim_{t \to +\infty} \hat{z}(t) = z(t).$$

Consequently, we can obtain the estimation of x by inverting the diffeomorphism $\phi(x)$. In summary, the following dynamics

$$\begin{aligned}\dot{\hat{z}} &= A_O\hat{z} + \beta(\bar{y}, w) + K_t(\varphi(y) - C_O\hat{z})\\ \dot{S} &= -\rho S - SA_O - A_O^T S + C_O^T C_O\\ K_t &= S^{-1}C_O^T\\ \hat{x} &= \phi^{-1}(\hat{z})\end{aligned} \tag{5.23}$$

is an asymptotic observer of (5.1).

5.4.2 Design Procedure

The following summarizes the procedure to design observer based on the proposed normal form (5.6).

Step 1: According to (3.2), calculate the associated 1-forms θ for (5.1);
Step 2: Determine the family of vector fields τ defined in (3.4) and (3.5);
Step 3: Extend system (5.1) with an auxiliary dynamics (5.7);
Step 4: Calculate the new family of vector fields σ according to (5.8) and (5.9);
Step 5: Calculate $l(w)$ and $\eta(y, w)$ according to the results stated in Remark 5.1;
Step 6: Compute $\bar{\Lambda}$ and $\bar{\omega}$ according to (5.13) and (5.14), and deduce the diffeomorphism $\phi(x)$ according to (5.17);
Step 7: Apply this diffeomorphism to calculate $\phi_*(f)$, with f being defined in (5.1);
Step 8: Design the proposed adaptive observer (5.23) to estimate x of (5.1).

Exercises

Exercise 5.1 Consider the following dynamical system:

$$\begin{cases}\dot{x}_1 = x_1x_3\\ \dot{x}_2 = x_1 + x_3^2\\ \dot{x}_3 = x_2\\ y = h(x) = x_3.\end{cases}$$

1. Compute the observability 1-forms θ_1, θ_2 and θ_3.
2. Compute the vector fields τ_1, τ_2 and τ_3.
3. Show that $[\tau_2, \tau_3] = \varphi(y)\tau_1 \neq 0$. What can we deduce from this fact?
4. Add an auxiliary dynamics $\dot{w} = -y$ and set $\sigma_1 = l(w)\tau_1$, then compute σ_2 and σ_3.
5. Determine $l(w)$ such that $[\sigma_2, \sigma_3] = 0$.

6. Find a new vector field σ_4 which is independent of and commute with σ_1, σ_2 and σ_3.
7. Set $\theta_4 = dw$ and compute the matrix $\bar{\Lambda} = (\theta_i(\sigma_j))$ to deduce the change of coordinates.
8. Show that this diffeomorphism can transform the studied system into the extended observer normal form (5.6).

Exercise 5.2 Consider the following dynamical system:

$$\begin{cases} \dot{x}_1 = x_1 \sin(x_3) - x_3^2 \\ \dot{x}_2 = x_1 \\ \dot{x}_3 = x_2 \\ y = h(x) = x_3. \end{cases}$$

Follow the same procedure as detailed in the above exercise to calculate the extended observer normal form.

References

1. Boutat, D.: Extended nonlinear observer normal forms for a class of nonlinear dynamical systems. Int. J. Robust Nonlinear Control **25**(3), 461–474 (2015)
2. Boutat, D., Busawon, K.: On the transformation of nonlinear dynamical systems into the extended nonlinear observable canonical form. Int. J. Control **84**(1), 94–106 (2011)
3. Busawon, K., Farza, M., Hammouri, H.: A simple observer for a class of nonlinear systems. Appl. Math. Lett. **11**(3), 27–31 (1998)
4. Califano, C., Moog, C.: The observer error linearization problem via dynamic compensation. IEEE Trans. Autom. Control **59**(9), 2502–2508 (2014)
5. Carroll, R., Lindorff, D.: An adaptive observer for single-input single-output linear systems. IEEE Trans. Autom. Control **18**(5), 428–435 (1973)
6. Gauthier, J.P., Hammouri, H., Othman, S.: A simple observer for nonlinear systems applications to bioreactors. IEEE Trans. Autom. Control **37**(6), 875–880 (1992)
7. Hammouri, H., de Leon Morales, J.: Observer synthesis for state-affine systems. In: Proceedings of the 29th IEEE Conference on Decision and Control, pp. 784–785 (1990)
8. Ioannou, P.A., Sun, J.: Robust Adaptive Control, vol. 1. PTR Prentice-Hall, Upper Saddle River (1996)
9. Tami, R., Boutat, D., Zheng, G.: Extended output depending normal form. Automatica **49**(7), 2192–2198 (2013)
10. Yu, L., Zheng, G., Boutat, D.: Adaptive observer for simultaneous state and parameter estimations for an output depending normal form. Asian J. Control **19**(1), 356–361 (2017)

Chapter 6
Output-Depending Observer Normal Form

Abstract In the previous three chapters, the desired observer normal forms (3.3), (4.3) and (5.6) have the common point, i.e., the linear part is a constant matrix of Brunovsky form and the nonlinear part is a function of known (measured) variables. In this chapter, we enlarge the desired observer normal form by allowing output-depending matrix for its linear part, and it is named as output-depending observer normal form [11, 12]. Like previous chapters, a set of geometric conditions will be deduced to guarantee the existence of a change of coordinates that brings the studied nonlinear dynamical system into the proposed output-depending observer normal form. As what has been done in the previous three chapters, the proposed algorithm enables us to seek a new frame of vector fields such that their Lie brackets are vanished. For such a normal form, this chapter synthesizes a non-asymptotic observer, called as step-by-step sliding mode observer, and the associated design procedure will be given at the end of this chapter.

6.1 Problem Statement

Consider the following single-output dynamical system

$$\begin{aligned} \dot{x} &= f(x) \\ y &= h(x), \end{aligned} \tag{6.1}$$

where $x \in \mathscr{X} \subseteq \mathbb{R}^n$ with $\mathscr{X}$ being a neighborhood of 0, and $y \in \mathbb{R}$. Without loss of generality it is assumed that f and h are smooth with $f(0) = 0$ and $h(0) = 0$.

We have shown in Chaps. 3–5 that such a nonlinear system might be transformed into the observer normal form with output injection (3.3), (4.3) or (5.6), either via a diffeomorphism on the state, on the output, or via the technique of immersion. However, there do exist certain systems for which all the mentioned techniques fail, i.e., the commutativity condition of Lie brackets cannot be satisfied. This can be shown by the following example.

D. Boutat and G. Zheng, *Observer Design for Nonlinear Dynamical Systems*,
Lecture Notes in Control and Information Sciences 487,
https://doi.org/10.1007/978-3-030-73742-9_6

Example 6.1 Consider the following dynamical system:

$$\begin{aligned} \dot{x}_1 &= \gamma(y) \\ \dot{x}_2 &= \alpha(y)x_1 \\ \dot{x}_3 &= x_2 \\ y &= x_3, \end{aligned} \tag{6.2}$$

where $\alpha(y) \neq 0$. Thus, following Definition 3.1, we obtain the following observability 1-forms:

$$\begin{aligned} \theta_1 &= dx_3 \\ \theta_2 &= dx_2 \\ \theta_3 &= \alpha(y)dx_1 + \alpha'(y)x_1 dx_3. \end{aligned}$$

Since $\alpha(y) \neq 0$, therefore those 1-forms are linearly independent, which implies that (6.2) is observable. Then, following (3.4) and (3.5), a straightforward calculation gives

$$\begin{aligned} \tau_1 &= \tfrac{1}{\alpha}\tfrac{\partial}{\partial x_1} \\ \tau_2 &= \tfrac{\partial}{\partial x_2} + x_2 \tfrac{\alpha'}{\alpha}\tau_1 \\ \tau_3 &= \tfrac{\partial}{\partial x_3} + x_2 \tfrac{\alpha'}{\alpha}\tau_2 - \left(\alpha x_1 \tfrac{\alpha'}{\alpha} + x_2^2 \left(\tfrac{\alpha'}{\alpha}\right)'\right)\tau_1. \end{aligned}$$

Based on the above vector fields, we can check that

$$\begin{aligned} [\tau_2, \tau_3] &= \tfrac{\alpha'}{\alpha}\tau_2 - x_2\left(\tfrac{\alpha'}{\alpha}\right)'\tau_1 - \left(x_2 \tfrac{\alpha'}{\alpha}\tfrac{\alpha'}{\alpha} + 2x_2\left(\tfrac{\alpha'}{\alpha}\right)'\right)\tau_1 \\ &= \tfrac{\alpha'}{\alpha}\tau_2 - \left(x_2\left(\tfrac{\alpha'}{\alpha}\right)^2 + 3x_2\left(\tfrac{\alpha'}{\alpha}\right)'\right)\tau_1, \end{aligned}$$

where $\alpha'(y) = \frac{d\alpha(y)}{dy}$.

According to Corollary 4.2, the above relationship suggests to seek a nowhere-vanishing function $l(y)$ from $[\tau_2, \tau_3]$. By applying the result stated in Corollary 4.2, we have

$$\mu_1(x) = -\left(x_2\left(\frac{\alpha'}{\alpha}\right)^2 + 3x_2\left(\frac{\alpha'}{\alpha}\right)'\right)$$

and

$$\mu_2(y) = \frac{\alpha'}{\alpha}.$$

Consequently, by solving $l^2\mu_2(y) - nll' = 0$, as stated in Corollary 4.2, we obtain

$$l(y) = e^{\frac{1}{n}\int_0^y \mu_2(s)ds} = \alpha^{\frac{1}{3}}.$$

With this $l(y)$, according to (4.7) and (4.8), we can calculate the new vector fields σ as follows:

$$\begin{aligned}\sigma_1 &= l\tau_1 \\ \sigma_2 &= l\tau_2 - x_2 l'\tau_1 \\ \sigma_3 &= l\tau_3 - 2x_2 l'\tau_2 + \left(\alpha x_1 l' + x_2^2 l''\right)\tau_1.\end{aligned}$$

It can be then checked that $[\sigma_2, \sigma_3]$ contains a term of the $\mu(x_2, y)\tau_1$ where

$$\mu(x_2, y) = l^2\mu_1(x) + \left(-x_2\alpha(l')^2 + x_2 l'' l + x_2\frac{\alpha'}{\alpha}l'l - 2x_2 l'^2\right).$$

As $\mu(x_2, y)$ depends on the variable $x_2 = \dot{y}$, thus the results presented in Chaps. 4 and 5 cannot be applied. For such a case, Chap. 5 provides a solution by introducing an auxiliary dynamics, which increases the dimension of the studied system. If we want to keep the same dimension without the introduction of auxiliary dynamics, an alternative is to modify the conventional structure of normal forms (3.3) and (4.3). In fact, if we take a look at system (6.2), we can notice that it can be written into the following form:

$$\begin{aligned}\dot{x} &= A(y)x + \beta(y) \\ y &= Cx\end{aligned} \tag{6.3}$$

with

$$A(y) = \begin{pmatrix} 0 & 0\,0 \\ \alpha(y) & 0\,0 \\ 0 & 1\,0 \end{pmatrix}$$

$$\beta(y) = [\gamma(y), 0, 0]^T$$

and

$$C = [0, 0, 1].$$

It can be checked that the pair $(A(y), C)$ is observable if $\alpha(y) \neq 0$, and different types of observers have already been proposed in the literature for such a special form, named as output-depending observer normal form since the matrix $A(y)$ is now a function of output. Therefore, it is natural to set this general form as a new target to relax the commutativity condition of Lie brackets, without augmenting the system's dimension. □

Motivated by the above example, this chapter presents how to seek a local diffeomorphism $z = \phi(x)$, such that the studied dynamical system (6.1) can be transformed into the following output-depending normal form:

$$\begin{aligned}\dot{z} &= A_O(y)z + \beta(y) \\ y &= C_O z,\end{aligned} \tag{6.4}$$

where the matrix $A_O(y)$ has the following form:

$$A_O(y) = \begin{pmatrix} 0 & \cdots & 0 & 0 & 0 \\ \alpha_1(y) & \cdots & 0 & 0 & 0 \\ \vdots & \ddots & \ddots & \cdots & \vdots \\ 0 & \cdots & \alpha_{n-2}(y) & 0 & 0 \\ 0 & \cdots & 0 & \alpha_{n-1}(y) & 0 \end{pmatrix}$$

and

$$C_O = \begin{pmatrix} 0 & \cdots & 0 & 1 \end{pmatrix}.$$

In general, the target normal form needs to be observable such that conventional observers proposed in the literature can be reused. Therefore, for the proposed output-depending normal form (6.4), it is assumed that $\alpha_i(y) \neq 0$ for $1 \leq i \leq n-1$ such that the observability rank condition is satisfied.

6.2 Analysis of the Output-Depending Normal Form

Before deducing the necessary and sufficient conditions to transform (6.1) into the proposed output-depending observer normal form (6.4), it is worth analyzing in which frame (6.4) is written and how to obtain such a frame using the Lie bracket.

Let us rewrite the vector field f that governs the dynamical system (6.4) as follows:

$$f = \beta_1 \frac{\partial}{\partial x_1} + \sum_{i=2}^{n} (\alpha_{i-1} x_{i-1} + \beta_i) \frac{\partial}{\partial x_i}.$$

Now, let τ_i for $1 \leq i \leq n$ be the vector fields associated with f which are defined in (3.4) and (3.5). It is clear that this frame does not commute, as shown in Example 6.1.

To build the correct frame which is commutative in the sense of Lie brackets, denote $\theta_i = dL_f^{i-1}h$ for $1 \leq i \leq n$ as the observability 1-forms, then according to (3.4), the first vector field τ_1 is

$$\tau_1 = \frac{1}{\alpha_1 \times \cdots \times \alpha_{n-1}} \frac{\partial}{\partial x_1}.$$

Hence, if we construct a new vector field σ_1 as

$$\sigma_1 = \alpha_1 \times \cdots \times \alpha_{n-1} \tau_1$$

and generate the following vector fields $\sigma_i = \frac{1}{\alpha_{i-1}}[\sigma_{i-1}, f]$ for $2 \leq i \leq n$, we can then have

$$\sigma_i = \frac{\partial}{\partial x_i}$$

for all $1 \leq i \leq n$. It is evident that the constructed new frame σ is commutative in the sense of Lie brackets, which is exactly the canonical basis where f is expressed.

The above analysis reveals a way how to construct a commutative frame σ_i from τ_i, provided that α_i is not equal to 0. The following will detail how to determine those nowhere-vanishing functions α_i which will be used to generate such a new frame σ_i.

6.3 Construction of New Vector Fields

Consider the dynamical system (6.1), according to Definition 3.1, let us denote

$$\theta = \begin{bmatrix} \theta_1 \\ \vdots \\ \theta_n \end{bmatrix},$$

with $\theta_i = dL_f^{i-1}h$ for $1 \leq i \leq n$. Since (6.1) is assumed to be observable in the sense that the observability rank condition is satisfied, thus θ_i for $1 \leq i \leq n$ are linearly independent. Note now the frame τ as

$$\tau = [\tau_1, \dots, \tau_n],$$

where τ_1 and τ_i for $2 \leq i \leq n$ are determined via (3.4) and (3.5).

Recall that if the commutativity conditions of Lie brackets fail to be fulfilled, then we want to seek a set of nowhere-vanishing functions $\alpha_i(y)$ for $1 \leq i \leq n-1$. For the sake of simplicity, denote

$$\pi_{n-1} = \alpha_1 \times \alpha_2 \times \cdots \times \alpha_{n-1} = \Pi_{i=1}^{n-1}\alpha_i \tag{6.5}$$

as the product of functions α_i for $1 \leq i \leq n-1$, and

$$\tau_1 = \frac{1}{\pi_{n-1}}\frac{\partial}{\partial x_1}.$$

To build the correct frame which is commutative in the sense of Lie brackets, define the first new vector field as

$$\sigma_1 = \pi_{n-1}\tau_1 \tag{6.6}$$

and calculate the following vector fields by induction:

$$\sigma_i = \frac{1}{\alpha_{i-1}}[\sigma_{i-1}, f]. \tag{6.7}$$

The key problem is how to determine those functions $\alpha_i(y)$. To achieve this goal, similar to the approaches used in the previous chapters, we need first to establish the relationship between the frame τ and the frame σ with $\sigma = [\sigma_1, \ldots, \sigma_n]$.

Lemma 6.1 *The vector fields of τ and the vector fields of σ are linked by the following equation:*

$$\begin{cases} \sigma_1 = \pi_{n-1}\tau_1 \\ \sigma_i = \sum_{j=1}^{i} a_{i,j}\tau_j, \text{ for } 2 \le i \le n, \end{cases} \tag{6.8}$$

where π_{n-1} is defined in (6.5), and $a_{i,k}$ are given by induction as follows:

$$\begin{cases} a_{i,1} = -\frac{1}{\alpha_{i-1}} L_f a_{i-1,1}, \text{ for } 2 \le i \le n \\ a_{i,k} = \frac{1}{\alpha_{i-1}}\left(-L_f a_{i-1,k} + a_{i-1,k-1}\right), \text{ for } 2 \le k \le i-1 \\ a_{i,i} = \alpha_{i+1} \times \cdots \times \alpha_{n-1} \text{ for } 1 \le i \le n-1 \\ a_{n,n} = 1. \end{cases} \tag{6.9}$$

Proof We will complete the proof by induction. Assume that (6.8) is true for the ith order, then the following will show that it is also true the $(i+1)$th order.

By definition, we know that $\sigma_{i+1} = \frac{1}{\alpha_i}[\sigma_i, f]$. Since (6.8) is satisfied for the ith order, thus we have

$$\sigma_{i+1} = \frac{1}{\alpha_i}\left[\sum_{j=1}^{i} a_{i,j}\tau_j, f\right].$$

Using the properties of Lie bracket and the fact $\tau_{i+1} = [\tau_i, f]$, we get

$$\sigma_{i+1} = \frac{1}{\alpha_i}\left(\sum_{j=1}^{i} - L_f\left(a_{i,j}\right)\tau_j + \sum_{j=1}^{i} a_{i,j}\tau_{j+1}\right).$$

By separating the first and the last terms, we obtain

$$\sigma_{i+1} = -\frac{1}{\alpha_i} L_f\left(a_{i,1}\right)\tau_1 + \frac{1}{\alpha_i}\left(\sum_{j=2}^{i} - L_f\left(a_{i,j}\right)\tau_j + \sum_{j=1}^{i-1} a_{i,j}\tau_{j+1}\right) + \frac{1}{\alpha_i} a_{i,i}\tau_{i+1}.$$

A change on the index j leads to

$$\sigma_{i+1} = -\frac{1}{\alpha_i} L_f\left(a_{i,1}\right)\tau_1 + \frac{1}{\alpha_i}\left(\sum_{j=2}^{i} - L_f\left(a_{i,j}\right)\tau_j + \sum_{j=2}^{i} a_{i,j-1}\tau_j\right) + \frac{a_{i,i}}{\alpha_i}\tau_{i+1}$$

and we proved that (6.8) is also true for the $(i+1)$th order. □

From (6.8) and (6.9), we can see that σ_i contains the term $L_f^k h$. Hence, it is necessary to know the coefficients of the highest Lie derivative $L_f^k h$ in each σ_i. To better understand this relation, let us take a look at the first three vector fields σ_i for $1 \leq i \leq 3$.

(1) For $i = 1$, according to (6.8), we have

$$\sigma_1 = \pi_{n-1}\tau_1.$$

Let us note the coefficient as $\gamma_{1,1} = \pi_{n-1}$ which is a function of the output y;

(2) For the second vector field, we have

$$\sigma_2 = -\frac{\pi'_{n-1}}{\alpha_1} L_f h \tau_1 + \frac{\pi_{n-1}}{\alpha_1}\tau_2,$$

and we can note $a_{2,1} = \gamma_{2,1} L_f h$ where the coefficient $\gamma_{2,1} = -\frac{\pi'_{n-1}}{\alpha_1}$ and $\gamma_{2,2} = \frac{\pi_{n-1}}{\alpha_1}$.

(3) For $i = 3$, we obtain

$$\sigma_3 = \frac{1}{\alpha_2}\left(\frac{\pi'_{n-1}}{\alpha_1} L_f^2 h + \left(\frac{\pi'_{n-1}}{\alpha_1}\right)' (L_f h)^2\right)\tau_1 - \frac{1}{\alpha_2}\left(\frac{\pi'_{n-1}}{\alpha_1} + \left(\frac{\pi_{n-1}}{\alpha_1}\right)'\right) L_f h \tau_2 + \frac{\pi_{n-1}}{\alpha_2\alpha_1}\tau_3.$$

In the function $a_{3,1} = \frac{1}{\alpha_2}\left(\frac{\pi'_{n-1}}{\alpha_1} L_f^2 h + \left(\frac{\pi'_{n-1}}{\alpha_1}\right)' (L_f h)^2\right)$, the coefficient of interest is $\gamma_{3,1} = \frac{\pi'_{n-1}}{\alpha_2\alpha_1}$ before the highest Lie derivative $L_f^2 h$, $a_{3,2} = \gamma_{3,2} L_f h$ with $\gamma_{3,2} = -\frac{1}{\alpha_2}\left(\frac{\pi'_{n-1}}{\alpha_1} + \left(\frac{\pi_{n-1}}{\alpha_1}\right)'\right)$, and $\gamma_{3,3} = \frac{\pi_{n-1}}{\alpha_2\alpha_1}$.

The above discussion shows a clue on how to relate α_i to σ_i through $\gamma_{i,i}$, which is the coefficient before the highest Lie derivative $L_f^k h$. This relation will be summarized by the following lemma.

Lemma 6.2 *Note*

$$\sigma_i = \sum_{j=1}^{i-1} \gamma_{i,j} L_f^{i-j} h \tau_j + \gamma_{i,i}\tau_i + R_i$$

for $4 \leq i \leq n$ *where* $\gamma_{i,j}$ *is the coefficient before* $L_f^{i-j} h$*, then those coefficients satisfy the following properties:*

(1) the coefficient before $L_f^i h$ *in the direction of vector field* τ_1 *equals*

$$\gamma_{i,1} = -\frac{\gamma_{i-1,1}}{\alpha_i};$$

(2) for $2 \leq j \leq i-2$*, the coefficient before* $L_f^{i-j} h$ *in the direction of* τ_j *satisfies*

$$\gamma_{i,j} = \frac{1}{\alpha_{i-1}}\left(\gamma_{i-1,j} - \gamma_{i-1,j-1}\right);$$

(3) *the coefficient before the Lie derivative $L_f h$ in the direction of τ_{i-1} fulfills*

$$\gamma_{i,i-1} = \frac{1}{\alpha_{i-1}}\left(\gamma_{i-1,i-2} - \gamma'_{i-1,i-1}\right);$$

(4) *for $j = i$, we have*

$$\gamma_{i,i} = \frac{\gamma_{i-1,i-1}}{\alpha_{i-1}}.$$

Proof According to (6.7), for $4 \le i \le n$, we have $\sigma_i = \frac{1}{\alpha_{i-1}}\left[\sigma_{i-1}, f\right]$. Therefore, write $\sigma_{i-1} = \sum_{j=1}^{i-1} a_{i-1,j}\tau_j$ as

$$\sigma_{i-1} = \sum_{j=1}^{i-1} \gamma_{i-1,j} L_f^{i-(j+1)} h\tau_j + R_{i-1},$$

where the R_{i-1} is a vector field that contains the low-order Lie derivatives of h. Note that the above equality is equivalent to the following one

$$\sigma_{i-1} = \sum_{j=1}^{i-2} \gamma_{i-1,j} L_f^{i-(j+1)} h\tau_j + \gamma_{i-1,i-1}\tau_{i-1} + R_{i-1}.$$

Then a straightforward calculation leads to

$$\begin{aligned}\sigma_i = & -\tfrac{\gamma_{i-1,1}}{\alpha_{i-1}} L_f^i h\tau_1 + \tfrac{1}{\alpha_{i-1}} \sum_{k=2}^{i-2} \left(\gamma_{i-1,k} - \gamma_{i-1,k-1}\right) L_f^{i-k} h\tau_k \\ & + \tfrac{1}{\alpha_{i-1}} \left(\gamma_{i-1,i-2} - \gamma'_{i-1,i-1}\right) L_f h\tau_{i-1} + \tfrac{\gamma_{i-1,i-1}}{\alpha_{i-1}}\tau_i + R_i,\end{aligned}$$

where R_i contains all the low-order Lie derivatives of h. Finally, we proved these 4 properties stated in Lemma 6.2. □

Recall that the objective to seek this new frame σ is that its vector fields should satisfy the commutativity condition of Lie brackets. Therefore, the following concerns the calculation of Lie brackets for σ.

Lemma 6.3 *The following statements hold:*

(1) $[\sigma_1, \sigma_i] = 0$ *for* $1 \le i \le n-1$ *if and only if* $[\tau_1, \tau_j] = 0$ *for* $1 \le j \le n-1$;
(2) $[\sigma_1, \sigma_n] = \left(\gamma_{1,1}\gamma_{1,n} - \gamma_{n,n}\gamma'_{1,1}\right)\tau_1 + \gamma_{1,1}\gamma_{n,n}[\tau_1, \tau_n]$.

Proof According to (3.4), we have $L_{\tau_1} L_f^i h = 0$ for $1 \le i \le n-2$ and $L_{\tau_1} L_f^{n-1} h = 1$. Since

$$\sigma_i = \sum_{j=1}^{i-1} \gamma_{i,j} L_f^{i-j} h\tau_j + \gamma_{i,i}\tau_i + R_i$$

for $1 \leq i \leq n-1$ where R_i contains low-order Lie derivatives of h, then we have

$$[\sigma_1, \sigma_i] = \sum_{j=1}^{i-1} \gamma_{i,j} L_f^{i-j} h \pi_{n-1}[\tau_1, \tau_j] + \gamma_{i,i} \pi_{n-1}[\tau_1, \tau_i] + [\tau_1, R_i].$$

Therefore, if we assume that the first item is true up to $i-1$, then $[\sigma_1, \sigma_i] = 0$ if and only if $[\tau_1, \tau_i] = 0$ because $[\tau_1, R_i] = 0$. Consequently, we have $[\sigma_1, \sigma_i] = 0$ for $1 \leq i \leq n-1$ if and only if $[\tau_1, \tau_j] = 0$ for $2 \leq j \leq i$. Also, a similar calculation can prove the item (2) of Lemma 6.3. □

According to Lemma 6.2, we have

$$\gamma_{n,n} = 1$$

$$\gamma_{1,1} = \pi_{n-1}$$

and

$$\gamma_{1,n} = (-1)^{n-1} \frac{\pi'_{n-1}}{\pi_{n-1}}.$$

Then, following the item (2) of Lemma 6.3, we have

$$[\sigma_1, \sigma_n] = -2\pi'_{n-1}\tau_1 + \pi_{n-1}[\tau_1, \tau_n],$$

if n is even, and

$$[\sigma_1, \sigma_n] = \pi_{n-1}[\tau_1, \tau_n],$$

if n is odd. It is clear that, for the case of odd dimension, we have $[\sigma_1, \sigma_n] = 0$ if and only if $[\tau_1, \tau_n] = 0$. However, for the case of even dimension, if we have $[\tau_1, \tau_n] = \mu(y)\tau_1$, then $[\sigma_1, \sigma_n] = 0$ is equivalent to the following differential equation

$$-2\pi'_{n-1} + \mu(y)\pi_{n-1} = 0.$$

The above equation leads to the following solution:

$$\pi_{n-1} = e^{\frac{1}{2}\int_0^y \mu(s)ds}.$$

Note that $\phi_{n-1} = \Pi_{i=1}^{n-1}\alpha_i(y)$ which contains $n-1$ variables to be determined, therefore, the above equation can solve the problem treated in this chapter for the case $n = 2$. In this case, we have

$$\alpha_1(y) = \phi_1 = e^{\frac{1}{2}\int_0^y \mu(s)ds}.$$

For the case $n \geq 3$, we need to seek more equations of $\alpha_i(y)$, and this can be obtained by calculating other Lie brackets than $[\tau_1, \tau_n]$. To clearly explain the idea, let us consider the simple case with $n = 3$. Since the dimension is odd, therefore, we must have $[\tau_1, \tau_3] = 0$ and we are wondering what we can get from $[\sigma_2, \sigma_3]$. A straightforward calculation gives

$$[\sigma_2, \sigma_3] = -\left(\frac{\alpha_1'}{\alpha_1} + 3\alpha_2'\right)\tau_2 + \alpha_2\,[\tau_2, \tau_3] + \nu(x)\tau_1.$$

In this case, we have two functions α_1 and α_2 to be determined. We would like to point out that the function $\alpha_2(y)$ is in fact given by a diffeomorphism on the output. Indeed, in the normal form

$$\begin{aligned}\dot{z}_1 &= \beta_1(y)\\ \dot{z}_2 &= \alpha_1(y)z_1 + \beta_2(y)\\ \dot{z}_3 &= \alpha_2(y)z_2 + \beta_3(y)\\ y &= z_3,\end{aligned}$$

if we set

$$\xi_3 = \bar{y} = \int_0^y \frac{ds}{\alpha_2(s)},$$

then we have

$$\dot{\xi}_3 = z_2 + \frac{\beta_3(y)}{\alpha_2(y)}.$$

The above fact means that we can always choose

$$\alpha_{n-1} = 1$$

for any dimensional nonlinear system. Consequently, we have in fact only $n - 2$ unknown functions $\alpha_i(y)$ for $1 \leq i \leq n - 2$ to be determined.

Applying the above remark to the case $n = 3$, $[\sigma_2, \sigma_3] = 0$ only if

$$[\tau_2, \tau_3] = \mu(y)\tau_2$$

with $\alpha_1(y)$ and $\mu(y)$ satisfying the following differential equation:

$$-\left(\frac{\alpha_1'}{\alpha_1} + 3\alpha_2'\right) + \alpha_2\mu(y) = 0,$$

where $\alpha_2 = 1$.

Hereafter, we assume that $\alpha_{n-1} = 1$ and we want to seek $n - 2$ unknown functions α_i for $1 \leq i \leq n - 2$. Note

$$\sigma_i = \sum_{j=1}^{i-1} \gamma_{i,j} L_f^{i-j} h\tau_j + \gamma_{i,i}\tau_i + R_i,$$

then the following provides a result to determine those $\alpha_i(y)$.

Lemma 6.4 *Assume that* $[\sigma_i, \sigma_s] = 0$ *for all* $1 \le s \le n-i$, *then*

(1) for $n-i+1 \le s \le n-1$, *we have*

$$[\sigma_i, \sigma_s] = \left(\gamma_{i,i}\gamma_{s,s+i-n} - \gamma_{s,s}\gamma_{i,s+i-n}\right)\tau_{s+i-n} + \gamma_{i,i}\gamma_{s,s}[\tau_i, \tau_s] + R_{i,s}; \quad (6.10)$$

(2) for $s = n$, *we have*

$$[\sigma_i, \sigma_n] = \left(\gamma_{i,i}\gamma_{n,i} - \gamma_{n,n}\gamma'_{i,i}\right)\tau_i + \gamma_{i,i}\gamma_{n,n}[\tau_i, \tau_n] + R_{i,n}. \quad (6.11)$$

Proof The proof of this lemma is similar to that of Lemma 6.3. □

The assumption imposed in Lemma 6.4 is consistent with the fact that $L_{\tau_i}L_f^s h = 0$ for $1 \le s \le n-i$. Another important fact is that, except $\gamma_{i,i}$, all other terms $\gamma_{i,j}$ for $i \ne j$ contain the derivatives of $\alpha_i(y)$.

Finally, $[\sigma_i, \sigma_s] = 0$ only if $[\tau_i, \tau_s] = \mu_{i,s}(y)\tau_{s+i-n} + R$ and the following equations:

$$\left(\gamma_{i,i}\gamma_{s,s+i-n} - \gamma_{s,s}\gamma_{i,s+i-n}\right) + \mu_{i,s}(y)\gamma_{i,i}\gamma_{s,s} = 0 \quad (6.12)$$

and

$$\left(\gamma_{i,i}\gamma_{n,i} - \gamma_{n,n}\gamma'_{i,i}\right) + \mu_{i,n}(y)\gamma_{i,i}\gamma_{n,n} = 0 \quad (6.13)$$

are satisfied, which enable us to deduce the unknown functions $\alpha_k(y)$ for $1 \le k \le n-2$.

After having determined the function $\alpha_i(y)$, we can then calculate the family of vector fields $\sigma = [\sigma_1, \ldots, \sigma_n]$ via (6.6) and (6.7). With the 1-forms $\theta = [\theta_1^T, \ldots, \theta_n^T]^T$ where $\theta_i = dL_f^{i-1}h$ for $1 \le i \le n$, we can then calculate the following matrix:

$$\bar{\Lambda} = \theta\sigma = \begin{pmatrix} 0 & 0 & \cdots & 0 & 1 \\ 0 & 0 & \cdots & \pi_{n-1}(y) & \bar{\Lambda}_{2,n} \\ \vdots & \vdots & \iddots & \vdots & \vdots \\ 0 & \pi_2(y) & \cdots & \bar{\Lambda}_{n-1,n-1} & \bar{\Lambda}_{n-1,n} \\ \pi_1(y) & \bar{\Lambda}_{n,2} & \cdots & \bar{\Lambda}_{n,n-1} & \bar{\Lambda}_{n,n} \end{pmatrix}. \quad (6.14)$$

It is clear that this matrix is invertible for nowhere-vanishing functions $\alpha_i(y)$ with $1 \le i \le n-1$, thus we can define the following 1-forms:

$$\bar{\omega} = \bar{\Lambda}^{-1}\theta := (\bar{\omega}_i)_{1 \le i \le n}. \quad (6.15)$$

Similar to previous chapters, we can state the following result that guarantees the change of coordinates.

Theorem 6.1 *There exists a local change of coordinates* $z = \phi(x)$ *that transforms dynamical system (6.1) into the nonlinear observer normal form (6.4) if and only* *if there exist nowhere-vanishing functions* $\alpha_i(y)$ *for* $1 \leq i \leq n-1$ *such that the* $[\sigma_i, \sigma_j] = 0$ *for all* $1 \leq i, \leq j \leq n$. *Furthermore, the change of coordinates is determined by*

$$z_i = \phi_i(x) = \int_\gamma \bar{\omega}_i, \tag{6.16}$$

where $\gamma : [0, 1] \to \mathscr{X}$ *is any smooth curve on contractile neighborhood of* 0 *such that* $\gamma(0) = 0$ *and* $\gamma(1) = x$.

Proof The proof of necessity is similar to that of Theorems 4.1 and 5.2. Therefore, the following gives only proof of sufficiency.

Suppose that such a change of coordinates $\phi(x)$ exists, then it will act on σ_i as follows:

$$\phi_*(\sigma_i) = \frac{\partial}{\partial z_i}.$$

To determine how f is transformed by means of ϕ_*, let us compute $\frac{\partial \phi_*(f)}{\partial z_i}$. As

$$\sigma_{i+1} = \frac{1}{\alpha_i}[\sigma_i, f]$$

for $1 \leq i \leq n-1$, then we have

$$\frac{\partial \phi_*(f)}{\partial z_i} = \left[\frac{\partial}{\partial z_i}, \phi_*(f)\right]$$

because $\frac{\partial}{\partial z_i}$ is a constant vector field. Therefore, we have

$$\frac{\partial \phi_*(f)}{\partial z_i} = \phi_*[\sigma_i, f] = \alpha_{i-1}\phi_*(\sigma_{i+1}) = \alpha_i \frac{\partial}{\partial z_{i+1}}$$

for $1 \leq i \leq n-1$. Thus, the component of $\phi_*(f)$ in the direction of $\frac{\partial}{\partial z_{i+1}}$ has the form

$$\alpha_i(y)\frac{\partial}{\partial z_{i+1}} + \beta(z_n)\frac{\partial}{\partial z_{i+1}}.$$

Now, if all the vector fields σ_i for $1 \leq i \leq n$ commute, then the change of coordinates is given by its differential as follows:

$$dz = \bar{\Lambda}^{-1}\theta,$$

where the components of the matrix $\bar{\Lambda}$ are given by (6.14). □

Example 6.2 Consider the following single output nonlinear dynamical system

$$\begin{aligned}\dot{x}_1 &= x_2 e^{x_3}\\ \dot{x}_2 &= e^{-2x_3} x_1\\ \dot{x}_3 &= x_2\\ y &= x_3.\end{aligned}$$

It is clear that the above system is observable because the observability 1-forms are as follows:

$$\begin{aligned}\theta_1 &= dx_3\\ \theta_2 &= dx_2\\ \theta_3 &= e^{-2x_3} dx_1 - 2e^{-2x_3} x_1 dx_3.\end{aligned}$$

A straightforward calculation shows that

$$\begin{aligned}\tau_1 &= e^{2x_3} \tfrac{\partial}{\partial x_1}\\ \tau_2 &= \tfrac{\partial}{\partial x_2} - 2x_2\tau_1\\ \tau_3 &= \tfrac{\partial}{\partial x_3} - 2x_2\tau_2 + \left(2e^{-2x_3} x_1 + e^{-x_3}\right)\tau_1.\end{aligned}$$

Thus, we have

$$[\tau_1, \tau_2] = [\tau_1, \tau_3] = 0$$

and

$$[\tau_2, \tau_3] = -2\frac{\partial}{\partial x_2} \neq 0.$$

Following the presented procedure, let us compute the new frame by setting

$$\sigma_1 = \alpha_1 \tau_1$$

then we get

$$\sigma_2 = \frac{1}{\alpha_1}[\sigma_1, f] = \tau_2 - \frac{\alpha_1' x_2}{\alpha_1}\tau_1$$

and

$$\sigma_3 = [\sigma_2, f] = \tau_3 - \frac{\alpha_1' x_2}{\alpha_1}\tau_2 + \left(\frac{\alpha_1'}{\alpha_1} e^{-2x_3} x_1 + \left(\frac{\alpha_1'}{\alpha_1}\right)' x_2^2\right)\tau_1.$$

The objective now is to seek $\alpha_1(y)$ such that $[\sigma_2, \sigma_3] = 0$. For this, we compute

$$[\sigma_2, \sigma_3] = [\tau_2, \tau_3] - \frac{\alpha_1'}{\alpha_1}\tau_2 + \left(-2x_2\frac{\alpha_1'}{\alpha_1} - x_2\left(\frac{\alpha_1'}{\alpha_1}\right)^2 + 2\left(\frac{\alpha_1'}{\alpha_1}\right)' x_2\right)\tau_1.$$

As $[\tau_2, \tau_3] = -2\tau_2$, we have

$$[\sigma_2, \sigma_3] = -\left(2 + \frac{\alpha_1'}{\alpha_1}\right)\tau_2 + \left(-2x_2\frac{\alpha_1'}{\alpha_1} - x_2\left(\frac{\alpha_1'}{\alpha_1}\right)^2 + 2\left(\frac{\alpha_1'}{\alpha_1}\right)' x_2\right)\tau_1.$$

In order to eliminate the τ_2 direction, we need to impose

$$2 + \frac{\alpha_1'}{\alpha_1} = 0,$$

which implies that $\alpha_1 = e^{-2y}$. Then we obtain

$$\sigma_1 = \frac{\partial}{\partial x_1}, \sigma_2 = \frac{\partial}{\partial x_2}$$

and

$$\sigma_3 = \frac{\partial}{\partial x_3} + e^{x_3}\frac{\partial}{\partial x_1}.$$

Now, it can be checked that

$$[\sigma_1, \sigma_2] = [\sigma_1, \sigma_3] = [\sigma_2, \sigma_3] = 0,$$

which implies that there exists a diffeomorphism to transform the studied system into the presented observer normal form (6.4).

In order to deduce such a diffeomorphism, let us compute

$$\Lambda = \begin{pmatrix} \theta_1(\sigma_1) & \theta_1(\sigma_2) & \theta_1(\sigma_3) \\ \theta_2(\sigma_1) & \theta_2(\sigma_2) & \theta_2(\sigma_3) \\ \theta_3(\sigma_1) & \theta_3(\sigma_2) & \theta_3(\sigma_3) \end{pmatrix} = \begin{pmatrix} 0 & 0 & 1 \\ 0 & 1 & 0 \\ e^{-2x_3} & 0 & e^{-x_3} - 2e^{-2x_3}x_1 \end{pmatrix}$$

whose inverse is

$$\Lambda^{-1} = \begin{pmatrix} \frac{1}{e^{-2x_3}}\left(-e^{-x_3} + 2x_1e^{-2x_3}\right) & 0 & \frac{1}{e^{-2x_3}} \\ 0 & 1 & 0 \\ 1 & 0 & 0 \end{pmatrix}.$$

Hence, we obtain the differential of the change of coordinates as follows:

$$\begin{pmatrix} dz_1 \\ dz_2 \\ dz_3 \end{pmatrix} = \Lambda^{-1}\begin{pmatrix} \theta_1 \\ \theta_2 \\ \theta_3 \end{pmatrix} = d\begin{pmatrix} x_1 + e^{x_3} \\ x_2 \\ x_3 \end{pmatrix},$$

which implies that the diffeomorphism is of the following form:

$$
\begin{aligned}
z_1 &= x_1 + e^{x_3} \\
z_2 &= x_2 \\
z_3 &= x_3
\end{aligned}
$$

with which the studied system can be transformed into

$$
\begin{aligned}
\dot{z}_1 &= 0 \\
\dot{z}_2 &= e^{-2y} z_1 - e^{-y} \\
\dot{z}_3 &= z_2 \\
y &= z_3.
\end{aligned}
$$

□

6.4 Observer Design

6.4.1 Step-by-Step Sliding Mode Observer

Compared to asymptotic observer, finite time one was less studied in the literature, which however is well appreciated in practice. Different methods have been proposed, such as sliding mode technique [6, 9], impulsive observer [3, 7], algebraic methods [1, 4]. This section presents a simple non-asymptotic observer for (6.1): step-by-step sliding mode observer [5, 10], based on the sliding mode technique.

For this, suppose that the state of (6.1) is bounded, so the state of (6.4) is also bounded. Then, let us rewrite (6.4) into the following form

$$
\begin{aligned}
\dot{z}_1 &= \beta_1(y) \\
\dot{z}_2 &= \alpha_1(y) z_1 + \beta_2(y) \\
&\vdots \\
\dot{z}_{n-1} &= \alpha_{n-2}(y) z_{n-2} + \beta_{n-1}(y) \\
\dot{z}_n &= \alpha_{n-1}(y) z_{n-1} + \beta_n(y) \\
y &= z_n.
\end{aligned}
\tag{6.17}
$$

Consider the following dynamics

$$
\begin{aligned}
\dot{\hat{z}}_1 &= \beta_1(y) + \lambda_1 \mathrm{sgn}(\tilde{z}_1 - \hat{z}_1) \\
\dot{\hat{z}}_2 &= \beta_2(y) + \lambda_2 \mathrm{sgn}(\tilde{z}_2 - \hat{z}_2) \\
&\vdots \\
\dot{\hat{z}}_{n-1} &= \beta_{n-1}(y) + \lambda_{n-1} \mathrm{sgn}(\tilde{z}_{n-1} - \hat{z}_{n-1}) \\
\dot{\hat{z}}_n &= \beta_n(y) + \lambda_n \mathrm{sgn}(y - \hat{z}_n)
\end{aligned}
\tag{6.18}
$$

with

$$\begin{aligned}
\tilde{z}_1 &= \tfrac{1}{\alpha_1(y)}\lambda_2 \mathrm{sgn}(\tilde{z}_2 - \hat{z}_2) \\
\tilde{z}_2 &= \tfrac{1}{\alpha_2(y)}\lambda_3 \mathrm{sgn}(\tilde{z}_3 - \hat{z}_3) \\
&\vdots \\
\tilde{z}_{n-2} &= \tfrac{1}{\alpha_{n-2}(y)}\lambda_{n-1} \mathrm{sgn}(\tilde{z}_{n-1} - \hat{z}_{n-1}) \\
\tilde{z}_{n-1} &= \tfrac{1}{\alpha_{n-1}(y)}\lambda_n \mathrm{sgn}(y - \hat{z}_n),
\end{aligned} \tag{6.19}$$

where $\lambda_i > 0$.

The step-by-step sliding mode observer works as follows. Firstly, consider the dynamics $\dot{z}_n$ and $\dot{\hat{z}}_n$, and note $e_i = z_n - \hat{z}_n$, we have

$$\dot{e}_n = \alpha_{n-1}(y)z_{n-1} - \lambda_n \mathrm{sgn}(y - \hat{z}_n). \tag{6.20}$$

Choose the following Lyapunov function

$$V_n = \frac{e_n^2}{2},$$

we get

$$\begin{aligned}
\dot{V}_n &= e_n\alpha_{n-1}(y)z_{n-1} - \lambda_n e_n \mathrm{sgn}(e_n) \\
&= e_n\alpha_{n-1}(y)z_{n-1} - \lambda_n |e_n|.
\end{aligned}$$

Since the term $\alpha_{n-1}(y)z_{n-1}$ is bounded, if we choose λ_n such that

$$\lambda_n > \sup_{t>0} \alpha_{n-1}(y(t))z_{n-1}(t),$$

then we have

$$\dot{V}_n < -\left[\lambda_n - \alpha_{n-1}(y)z_{n-1}\right]\sqrt{2}V_n^{\frac{1}{2}},$$

which implies that V_n will converge to 0 in a finite time. Consequently, $\hat{z}_n$ will converge to z_n in a finite time.

Suppose that when $t > T_1$, we have $\hat{z}_n = z_n$, thus $e_n = 0$ and $\dot{e}_n = 0$. According to (6.20), we obtain the exact estimation of z_{n-1}, noted as $\tilde{z}_{n-1}$ which has the following form:

$$z_{n-1} = \frac{1}{\alpha_{n-1}(y)}\lambda_n \mathrm{sgn}(y - \hat{z}_n), \forall t > T_1.$$

According to (6.19), we can then deduce that

$$\tilde{z}_{n-1} = z_{n-1}, \forall t > T_1$$

and

$$\begin{aligned}
\dot{\hat{z}}_{n-1} &= \beta_{n-1}(y) + \lambda_{n-1}\mathrm{sgn}(\tilde{z}_{n-1} - \hat{z}_{n-1}) \\
&= \beta_{n-1}(y) + \lambda_{n-1}\mathrm{sgn}(z_{n-1} - \hat{z}_{n-1})
\end{aligned}, \quad \forall t > T_1.$$

The next step is then to define $e_{n-1} = z_{n-1} - \hat{z}_{n-1}$, and $V_{n-1} = \frac{e_{n-1}^2}{2}$. By choosing

$$\lambda_{n-1} > \sup_{t>0} \alpha_{n-2}(y(t))z_{n-2}(t)$$

similarly, we can prove that V_{n-1} converges to 0 in a finite time. Therefore, there exists $T_2 > T_1 > 0$ such that

$$\hat{z}_{n-1} = \tilde{z}_{n-1} = z_{n-1}, \forall t > T_2.$$

By iterating the same procedure, it is easy to prove that, there exist $T_n > \cdots > T_1$ such that

$$\hat{z}_i = z_i, \forall t > T_n$$

for $1 \leq i \leq n$. From the procedure of the proof, it is clear that the observation errors converge to 0 in a step-by-step way, i.e., e_i converges to 0 in a finite time before e_{i-1}. For this reason, the proposed observer is named as a step-by-step sliding mode observer.

Finally, we can obtain the estimation of x by inverting the diffeomorphism $\phi(x)$. In summary, the following dynamics:

$$\begin{aligned} \dot{\hat{z}}_1 &= \beta_1(y) + \lambda_1 \mathrm{sgn}(\tilde{z}_1 - \hat{z}_1) \\ \dot{\hat{z}}_2 &= \beta_2(y) + \lambda_2 \mathrm{sgn}(\tilde{z}_2 - \hat{z}_2) \\ &\vdots \\ \dot{\hat{z}}_{n-1} &= \beta_{n-1}(y) + \lambda_{n-1} \mathrm{sgn}(\tilde{z}_{n-1} - \hat{z}_{n-1}) \\ \dot{\hat{z}}_n &= \beta_n(y) + \lambda_n \mathrm{sgn}(y - \hat{z}_n) \\ \hat{x} &= \phi^{-1}(\hat{z}), \end{aligned} \tag{6.21}$$

where $\tilde{z}_i$ is defined in (6.19), is a non-asymptotic observer of (6.1).

6.4.2 Design Procedure

The following summarizes the procedure to design observer based on the proposed normal form (6.4).

Step 1: According to (3.2), calculate the associated 1-forms θ for (6.1);

Step 2: Determine the family of vector fields τ defined in (3.4) and (3.5);

Step 3: Calculate the new family of vector fields σ according to (6.6) and (6.7);

Step 4: Deduce $\alpha_k(y)$ for $1 \leq k \leq n-2$ according to (6.12) and (6.13);

Step 5: Compute $\bar{A}$ and $\bar{\omega}$ according to (6.14) and (6.15), then deduce the diffeomorphism $\phi(x)$ according to (6.16);

Step 6: Apply this diffeomorphism to calculate $\phi_*(f)$, with f being defined in (6.1);

Step 7: Design the proposed observer (6.21) to estimate x of (6.1).

Exercises

Exercise 6.1 Consider the following dynamical system:

$$\begin{cases} \dot{x}_1 = x_1x_2 - x_1^2 \\ \dot{x}_2 = x_1 \\ y = h(x) = x_2. \end{cases}$$

Find its output-depending observer normal form.

Exercise 6.2 Consider the following output depending observer normal form:

$$\begin{cases} \dot{x}_1 = \beta_1(y) \\ \dot{x}_2 = x_1 + \beta_2(y) \\ \dot{x}_3 = \alpha(y)x_2 + \beta_3(y) \\ y = h(x) = x_3. \end{cases}$$

Show that it can be transformed into an observer normal form.

Exercise 6.3 Consider the nonlinear dynamical system described by the following equations [2, 8]

$$\begin{cases} \dot{\chi}_1 = \varphi_{1,1}(y)\chi_1 + \psi_{1,n-1}(y)\chi_{n-1} + \varphi_{1,n}(y) \\ \dot{\chi}_i = \varphi_{i,1}(y)\chi_{i-1} + \varphi_{i,i}(y)\chi_i + \psi_{i,n-1}(y)\chi_{n-1} + \varphi_{i,n}(y) \quad \text{for } 2 \leq i \leq n-3 \\ \dot{\chi}_{n-2} = \varphi_{n-2,1}(y)\chi_{n-3} + \varphi^1_{n-2,n-2}(y)\chi_{n-2} + \psi_{n-2,n-1}(y)\chi_{n-1} \\ \qquad +\psi^1_{n-2,n-1}(y)\chi^2_{n-1} + \psi_{n-2,n}(y) \\ \dot{\chi}_{n-1} = \psi_{n-1,1}(y)\chi_{n-2} + \psi_{n-1,n-1}(y)\chi_{n-1} + \psi^1_{n-1,n-1}(y)\chi^2_{n-1} + \psi_{n-1,n}(y) \\ \dot{\chi}_n = \psi_{n,1}(y)\chi_{n-1} + \varphi_{n,n}(y) \\ y = \chi_n, \end{cases} \tag{6.22}$$

where $\chi = (\chi_1, \ldots, \chi_n)^T \in U \subset \mathbb{R}^n$ represents the state in a neighborhood of 0.

It is assumed that dynamical system (6.22) satisfies the observability rank condition, i.e., for $2 \leq i \leq n-2$ we have $\varphi_{i,1}(y) \neq 0$, $\psi_{n-1,1}(y) \neq 0$ and $\psi_{n,1}(y) \neq 0$ for all y. Seek a change of coordinates together with an auxiliary dynamics that transforms the above dynamical system into the following observer normal form:

$$\begin{aligned} \dot{x} &= A(y, w)x + \beta(y, w) \\ y &= z_n, \end{aligned}$$

where $A(y, w)$ is a matrix of Brunovsky canonical form, depending on the output and the auxiliary variables.

Exercise 6.4 Consider the following affine dynamical system:

$$\begin{cases} \dot{x}_1 = x_1 x_2 + x_1 u \\ \dot{x}_2 = x_1 \\ y = x_2. \end{cases}$$

By increasing the dimension of the above dynamical system with the auxiliary dynamics $\dot{w} = -u$, show that the following change of coordinates

$$\begin{aligned} z_1 &= -\tfrac{1}{2}e^{w}x_2^2 + x_1 e^{w} \\ z_2 &= e^{w}x_2 \end{aligned}$$

will transform the studied system into the following observer normal form:

$$\begin{cases} \dot{z}_1 = \tfrac{1}{2}e^{-w}y_2^2 u \\ \dot{z}_2 = z_1 + \tfrac{1}{2}e^{-w}y_2^2 - ye^{w}u \\ \dot{w} = -u \\ y = z_2. \end{cases}$$

References

1. Barbot, J.P., Fliess, M., Floquet, T.: An algebraic framework for the design of nonlinear observers with unknown inputs. In: Proceedings of the 46th IEEE Conference on Decision and Control, pp. 384–389 (2007)
2. Boutat, D.: Extended nonlinear observer normal forms for a class of nonlinear dynamical systems. Int. J. Robust Nonlinear Control **25**(3), 461–474 (2015)
3. Engel, R., Kreisselmeier, G.: A continuous-time observer which converges in finite time. IEEE Trans. Autom. Control **47**(7), 1202–1204 (2002)
4. Fliess, M., Sira-Ramırez, H.: Reconstructeurs d'état. Comptes Rendus Mathematique **338**(1), 91–96 (2004)
5. Floquet, T., Barbot, J.P., Perruquetti, W.: Higher-order sliding mode stabilization for a class of nonholonomic perturbed systems. Automatica **39**(6), 1077–1083 (2003)
6. Fridman, L., Levant, A., Davila, J.: Observation of linear systems with unknown inputs via high-order sliding-modes. Int. J. Syst. Sci. **38**(10), 773–791 (2007)
7. Raff, T., Allgower, F.: An impulsive observer that estimates the exact state of a linear continuous-time system in predetermined finite time. In: Proceedings of Mediterranean Conference on Control and Automation, pp. 1–3 (2007)
8. Tami, R., Boutat, D., Zheng, G.: Extended output depending normal form. Automatica **49**(7), 2192–2198 (2013)
9. Utkin, V.: Sliding Modes in Control and Optimization. Springer Science & Business Media, New York (2013)
10. Xiong, Y., Saif, M.: Sliding mode observer for nonlinear uncertain systems. IEEE Trans. Autom. Control **46**(12), 2012–2017 (2001)

11. Zheng, G., Boutat, D., Barbot, J.P.: Output dependent observability linear normal form. In: Proceedings of the 44th IEEE Conference on Decision and Control and European Control Conference, pp. 7026–7030 (2005)
12. Zheng, G., Boutat, D., Barbot, J.P.: Single output-dependent observability normal form. SIAM J. Control Optim. **46**(6), 2242–2255 (2007)

Chapter 7
Extension to Nonlinear Partially Observable Dynamical Systems

Abstract In the previous chapters, we have applied differential geometric method to design observers for nonlinear dynamical systems whose states are supposed to be fully observable. However, in practice some systems might be only partially observable, i.e., only a part of states are observable. For this topic, several works have been done by using different techniques, such as reduced-order observer and LMI technique [12], differential geometric method [3, 7] where the desired normal form is quite special. The objective of this chapter is to show how to treat general partial observable case when applying differential geometric method. In this chapter, we will use the notion of commutativity of Lie bracket modulo a distribution, and at the same time apply a diffeomorphism on the output [10, 11]. Also, necessary and sufficient geometric conditions have been established to guarantee the existence of a diffeomorphism such that the studied nonlinear system can be transformed into the desired general partial observer normal form. A homogeneous finite-time observer is then designed for such a normal form at the end of this chapter.

7.1 Problem Statement

Consider the following nonlinear dynamical system

$$\begin{aligned} \dot{x} &= f(x) \\ y &= h(x), \end{aligned} \tag{7.1}$$

where $x \in \mathscr{X} \subset \mathbb{R}^n$ with $\mathscr{X}$ being a neighborhood of 0, and $y \in \mathbb{R}$. Without loss of generality it is assumed that f and h are smooth with $f(0) = 0$ and $h(0) = 0$.

For system (7.1), as we have presented in Chap. 3, if the pair $(f(x), h(x))$ locally satisfies the observability rank condition on $\mathscr{X}$, i.e.,

$$\text{rank}\left\{dh, dL_f h, \ldots, dL_f^n h\right\}(x) = n,$$

for $x \in \mathscr{X}$, then the following 1-forms

D. Boutat and G. Zheng, *Observer Design for Nonlinear Dynamical Systems*, Lecture Notes in Control and Information Sciences 487,
https://doi.org/10.1007/978-3-030-73742-9_7

$$\theta_1 = dh \text{ and } \theta_i = dL_f^{i-1}h, \text{ for } 2 \leq i \leq n$$

are independent on $\mathscr{X}$. Therefore, there exists a family of vector fields

$$\tau = [\tau_1, \ldots, \tau_n],$$

where the first vector field τ_1 is the solution of (3.4) and other vector fields are obtained by (3.5). According to Theorem 3.2, if

$$\left[\tau_i, \tau_j\right] = 0 \text{ for } 1 \leq i, j \leq n, \tag{7.2}$$

then system (7.1) can be transformed, by means of a local change of coordinates $z = \phi(x)$, into the nonlinear observer normal form (3.3).

Obviously, for the nonlinear dynamical system (7.1), if

$$\text{rank}\left\{dh, dL_f h, \ldots, dL_f^n h\right\}(x) = r < n$$

for $x \in \mathscr{X}$, which implies that only a part of states of the studied system are observable, the proposed method in Chap. 3 could not be applied.

This chapter will deal with the geometric conditions that enable us to transform a partial observable dynamical system into the following normal form:

$$\begin{aligned} \dot{z} &= A_O z + \beta(y) \\ \dot{\zeta} &= \eta(z, \zeta) \\ y &= C_O z, \end{aligned} \tag{7.3}$$

where $z \in \mathbb{R}^r$, $\zeta \in \mathbb{R}^{n-r}$, $y \in \mathbb{R}$, $\beta : \mathbb{R} \to \mathbb{R}^r$, $\eta : \mathbb{R}^r \times \mathbb{R}^{n-r} \to \mathbb{R}^{n-r}$, and

$$C_O = (0, \ldots, 0, 1) \in \mathbb{R}^{1\times r}$$

and A_O is the $r \times r$ Brunovsky matrix

$$A_O = \begin{pmatrix} 0 & 0 & \cdots & 0 & 0 \\ 1 & 0 & \ddots & \vdots & \vdots \\ 0 & 1 & \cdots & 0 & 0 \\ 0 & \cdots & \cdots & 0 & 0 \\ 0 & \cdots & \cdots & 1 & 0 \end{pmatrix} \in \mathbb{R}^{r\times r}.$$

For (7.3), we can easily design different types of observers to estimate the part of observable state z, which will be presented in Sect. 7.4. The following will discuss how to deduce a change of coordinates that transforms (7.1) into the target normal form (7.3).

7.2 Necessary and Sufficient Conditions

As we require only the partial observability for the studied system (7.1), thus we assume that

$$\text{rank}\left\{dh, dL_f h, \dots, dL_f^n h\right\}(x) = r < n$$

for $x \in \mathscr{X} \subset \mathbb{R}^n$.

As for the previous chapters of this book, denote the observability 1-forms for $1 \leq i \leq r$ by $\theta_i = dL_f^{i-1}h$ and note

$$\Delta = \text{span}\{\theta_1, \theta_2, \dots, \theta_r\}$$

as the co-distribution spanned by the observability 1-forms. The partial observability rank condition implies that

$$dL_f^i h \in \Delta$$

for $i \geq r + 1$. In the same way as for the full observability, we define the following distribution $\Delta^{\perp}$ as follows:

$$\Delta^{\perp} = \ker\Delta = \{X : \theta_k(X) = 0, \text{ for } 1 \leq k \leq r\}. \tag{7.4}$$

The following result provides some useful properties of distribution $\Delta^{\perp}$ defined in (7.4) that will be used within this chapter.

Lemma 7.1 *The distribution $\Delta^{\perp}$ has the following properties:*

C1: $\Delta^{\perp}$ is involutive in the sense that for any two vector fields $H_1 \in \Delta^{\perp}$ and $H_2 \in \Delta^{\perp}$, we have

$$[H_1, H_2] \in \Delta^{\perp}.$$

C2: there exist $(n - r)$ vector fields $\{\bar{\tau}_{r+1}, \dots, \bar{\tau}_n\}$ that span $\Delta^{\perp}$ such that

$$[\bar{\tau}_i, \bar{\tau}_j] = 0.$$

for $r + 1 \leq i, j \leq n$, i.e., $\{\bar{\tau}_{r+1}, \dots, \bar{\tau}_n\}$ is the commutative basis of $\Delta^{\perp}$;

C3: $\Delta^{\perp}$ is f-invariant in the sense that for any vector field $H \in \Delta^{\perp}$, we have

$$[f, H] \in \Delta^{\perp}.$$

Proof Let us recall from (2.16) the following equality:

$$d\theta_i(H_1, H_2) = L_{H_1}\theta_i(H_2) - L_{H_2}\theta_i(H_1) - \theta_i\left([H_1, H_2]\right). \tag{7.5}$$

Since $\theta_i = dL_f^{i-1}h$ for $1 \leq i \leq r$, then they are exact 1-forms (thus closed), and the above formula becomes

$$\theta_i\left([H_1, H_2]\right) = L_{H_1}\theta_i(H_2) - L_{H_2}\theta_i(H_1). \tag{7.6}$$

Now, as $H_1 \in \Delta^\perp$ and $H_2 \in \Delta^\perp$, we have

$$[H_1, H_2] \in \Delta^\perp$$

because $\theta_i\left([H_1, H_2]\right) = 0$. Thus $\Delta^\perp$ is involutive, and C1 is fulfilled in Lemma 7.1.

In regard of the second property C2 in Lemma 7.1, it is a consequence of C1 thanks to the well-known Frobenius' Lemma 2.3 stated in Chap. 2.

In order to prove the third property C3 in Lemma 7.1, we need to show that $[f, H] \in \Delta^\perp$. For this, we will use again (7.6) by replacing H_1 by f and H_2 by H. As θ_i for $1 \leq i \leq r$ are exact, then according to the Eq. (7.6), we have

$$\theta_i\left([f, H]\right) = L_f\theta_i(H) - L_H\theta_i(f).$$

Since $H \in \Delta^\perp$, then $\theta_i(H) = 0$ which implies that

$$L_f\theta_i(H) = 0.$$

Therefore, $\theta_i\left([f, H]\right) = 0$ is equivalent to $L_H\theta_i(f) = 0$. Moreover, based on the Cartan's identity given in (2.13) and the fact that $d\theta_i(f) = \theta_{i+1}$ for $1 \leq i \leq r-1$, we have

$$L_H\theta_i(f) = d\imath_H\theta_i(f) + \imath_H d\theta_i(f) = \theta_{i+1}(H) = 0,$$

which implies that $[f, H] \in \Delta^\perp$. Therefore $\Delta^\perp$ is f-invariant. □

Now, if τ_1 is one of the vector fields that fulfills the following under-determined algebraic equations:

$$\begin{aligned} \theta_k(\tau_1) &= 0 \text{ for } 1 \leq k \leq r-1 \\ \theta_r(\tau_1) &= 1, \end{aligned} \tag{7.7}$$

then, for any $H \in \Delta^\perp$, $\tau_1 + H$ is also a solution of (7.7). As in Chap. 3, we build by induction a family of vector fields from any chosen τ_1 satisfying (7.7) as follows:

$$\tau_i = \left[\tau_{i-1}, f\right] \text{ for } 2 \leq i \leq r. \tag{7.8}$$

Let $\tilde{\tau}_1$ be another solution of (7.7) and construct another family of $(r-1)$ vector fields

$$\tilde{\tau}_i = \left[\tilde{\tau}_{i-1}, f\right] \text{ for } 2 \leq i \leq r. \tag{7.9}$$

The following lemma provides the relationship between the above two families of vector fields for the different chosen τ_1 and $\tilde{\tau}_1$.

Lemma 7.2 *There exists a family of $H_i \in \Delta^\perp$ for $1 \leq i \leq r$ such that*

$$\tilde{\tau}_i = \tau_i + H_i,$$

where τ_i and $\tilde{\tau}_i$ are defined in (7.8) and (7.9), respectively.

Proof Let us prove it for $\tilde{\tau}_2$ and the same method works for the rest of the vector fields. For this, we have

$$\begin{aligned}\tilde{\tau}_2 &= [\tilde{\tau}_1, f] = [\tau_1 + H_1, f] \\ &= [\tau_1, f] + [H_1, f] = \tau_2 + H_2\end{aligned}$$

in the above we used the property that $\Delta^\perp$ is f-invariant, i.e., $[H_1, f] = H_2 \in \Delta^\perp$.

Then, by induction, we can deduce that for $1 \le i \le r$ there exists $H_i \in \Delta^\perp$ such that $\tilde{\tau}_i = \tau_i + H_i$. □

The above Lemma 7.2 stated that the family of vector fields τ_i for $1 \le i \le r$ are defined modulo $\Delta^\perp$. The next result shows other important properties of vector fields τ_i.

Lemma 7.3 *Given a solution τ_1 of (7.7) based on which a family of τ_i for $2 \le i \le r$ are obtained from (7.8), then it satisfies the following properties:*

(1) $\Delta^\perp$ is τ_i-invariant, thus for any $H \in \Delta^\perp$, we have $[\tau_i, H] \in \Delta^\perp$;
(2) The Lie bracket of vector fields of the family of τ_i is defined modulo $\Delta^\perp$, thus

$$[\tau_i + H_i, \tau_j + H_j] = [\tau_i, \tau_j] \text{ modulo } \Delta^\perp;$$

for any $H_i \in \Delta^\perp$ and $H_j \in \Delta^\perp$ with $1 \le i, j \le r$, we have
(3) For $1 \le i \le r$ and $r + 1 \le j \le n$, we have

$$\left[\tau_i, \bar{\tau}_j\right] \in \Delta^\perp, \tag{7.10}$$

where $\bar{\tau}_j$ was defined in C2 of Lemma 7.1, which is the commutative basis of $\Delta^\perp$.

Proof (1) Note that $d\theta_k = 0$ since $\theta_k = dL_f^{k-1}h$ for $1 \le k \le r$ is closed, then for $H \in \Delta^\perp$, we obtain

$$\begin{aligned}\theta_k\left([\tau_i, H]\right) &= L_{\tau_i}\theta_k(H) - L_H\theta_k(\tau_i) \\ &= -L_H\theta_k(\tau_i),\end{aligned}$$

where we used the fact that $H \in \Delta^\perp$, then it belongs to the kernel of θ_k.

According to the definition of τ_1, then the function $\theta_k(\tau_1)$ is constant, thus its Lie derivative is 0. Therefore, we obtain

$$\theta_k\left([\tau_1, H]\right) = -L_H\theta_k(\tau_1) = 0,$$

which yields $[\tau_1, H] \in \Delta^\perp$.

Now, set $i = 2$ and use Jacobi's identity defined in (2.15) to get

$$[\tau_2, H] = [[\tau_1, f], H] = -[[H, \tau_1], f] - [[H, f], \tau_1].$$

Since $\Delta^\perp$ is f and τ_i-invariant we obtain $[\tau_2, H] \in \Delta^\perp$.

Using repeatably Jacobi's identity, the invariance and the involutivity of $\Delta^\perp$, by induction we get the following result:

$$[\tau_i + H_i, \tau_j + H_j] = [\tau_i, \tau_j] \text{ modulo } \Delta^\perp.$$

(3) Due to the fact that $\Delta^\perp$ is τ_i-invariant and the fact that $\bar{\tau}_j$ for $r+1 \le j \le n$ are the basis of $\Delta^\perp$, we obtain

$$\left[\tau_i, \bar{\tau}_j\right] \in \Delta^\perp.$$

□

Lemma 7.4 *For a given τ_i for $2 \le i \le r$ there exists a family of vector fields $\bar{\tau}_i = \tau_i$ modulo $\Delta^\perp$ such that $\left[\bar{\tau}_i, \bar{\tau}_j\right] = 0$ for $1 \le i, j \le r$, if and only if $[\tau_i, \tau_j] \in \Delta^\perp$.*

Proof **Necessity:** Suppose that we have $\bar{\tau}_i = \tau_i + H_i, \bar{\tau}_j = \tau_j + H_j$ for $1 \le i, j \le r$ and $H_i \in \Delta^\perp$ and $H_j \in \Delta^\perp$ such that $\left[\bar{\tau}_i, \bar{\tau}_j\right] = 0$, then from the second item of Lemma 7.3, we obtain

$$0 = \left[\bar{\tau}_i, \bar{\tau}_j\right] = [\tau_i, \tau_j] \text{ modulo } \Delta^\perp,$$

which leads to $[\tau_i, \tau_j] \in \Delta^\perp$.

Sufficiency: Let us assume that $[\tau_i, \tau_j] \in \Delta^\perp$ for $1 \le i, j \le r$. We will prove that there exists $H_i \in \Delta^\perp$ for $1 \le i \le r$ such that, for $\bar{\tau}_k = \tau_k + H_k$,we have $[\bar{\tau}_i, \bar{\tau}_j] = 0$.

For this, let us consider a family of vector fields $\{\bar{\tau}_{r+1}, \dots, \bar{\tau}_n\}$ that is a commutative basis of $\Delta^\perp$. It has been stated in [5] that we can complete this basis by r independent commutative vector fields $\{X_1, \dots, X_r\}$, such that they commute with $\{\bar{\tau}_{r+1}, \dots, \bar{\tau}_n\}$ and form together a basis of the tangent fiber bundle $T\mathscr{X}$ of $\mathscr{X}$.

Then, on this basis, the vector fields τ_i $1 \le i \le r$ can be expressed as follows:

$$\tau_i = \sum_{l=1}^{r} \mu_{i,l}(x) X_l + \sum_{l=r+1}^{n} \mu_{i,l}(x) \bar{\tau}_l.$$

By setting

$$H_i = -\sum_{l=r+1}^{n} \mu_{i,l}(x) \bar{\tau}_l \in \Delta^\perp$$

and

$$H_j = \sum_{l=r+1}^{n} \mu_{j,l}(x) \bar{\tau}_l \in \Delta^\perp,$$

we obtain

$$\bar{\tau}_i = \tau_i + H_i = \sum_{l=1}^{r} \mu_{i,l}(x) X_l$$

and

$$\bar{\tau}_j = \tau_j + H_j = \sum_{l=1}^{r} \mu_{j,l}(x) X_l.$$

Now, from Lemma 7.3, we know that

$$[\tau_i, \tau_j] = [\bar{\tau}_i, \bar{\tau}_j] \text{ modulo } \Delta^{\perp}. \tag{7.11}$$

As we have $[X_k, X_s] = 0$ for $1 \leq k, s \leq r$, then Eq. (7.11) becomes

$$[\tau_i, \tau_j] = \left[\sum_{l,s=1}^{r} \mu_{i,l} L_{X_l} \mu_{j,s} X_s - \sum_{s,l=1}^{r} \mu_{j,s} L_{X_s} \mu_{i,l} X_l \right] \text{ modulo } \Delta^{\perp}.$$

Therefore, if $[\tau_i, \tau_j] \in \Delta^{\perp}$, then, there exist $H_i \in \Delta^{\perp}$ and $H_j \in \Delta^{\perp}$ defined above, such that

$$[\bar{\tau}_i, \bar{\tau}_j] = \sum_{l,s=1}^{r} \mu_{i,l} L_{X_l} \mu_{j,s} X_s - \sum_{s,l=1}^{r} \mu_{j,s} L_{X_s} \mu_{i,l} X_l = 0$$

thanks to the fact that $X_k \notin \Delta^{\perp}$. □

Let us provide an example to highlight the above results.

Example 7.1 Consider the following dynamical system

$$\begin{aligned} \dot{z}_1 &= 0 \\ \dot{z}_2 &= z_1 + y^2 \\ \dot{\eta} &= -\eta + \nu z_1^2 + y \\ y &= z_2. \end{aligned} \tag{7.12}$$

A straightforward calculation gives the following observability 1-forms

$$\begin{aligned} \theta_1 &= dz_2 \\ \theta_2 &= dz_1 + 2z_2 dz_2. \end{aligned}$$

The intersection of their kernel defines the following co-distribution

$$\Delta = \text{span}\{\theta_1, \theta_2\}$$

and the associated distribution

$$\Delta^{\perp} = \text{span}\left\{\frac{\partial}{\partial \eta}\right\}.$$

From (7.7), we obtain

$$\tau_1 = \frac{\partial}{\partial z_1} + r_1(z_1, z_2, \eta)\frac{\partial}{\partial \eta},$$

where $r_1(z_1, z_2, \eta)$ is a function of its arguments. From (7.8), we get

$$\tau_2 = [\tau_1, f] = \frac{\partial}{\partial z_2} + r_2\frac{\partial}{\partial \eta},$$

where $r_2 = 2\mu z_1 - r_1 - \frac{\partial r_1}{\partial z_2}(z_1 + z_2^2) - \frac{\partial r_1}{\partial \eta}(-\eta + \nu z_1^2 + z_2)$. The Lie bracket of the above two vector fields leads to

$$[\tau_1, \tau_2] = \left[\frac{\partial r_2}{\partial z_1} + \frac{\partial r_2}{\partial \eta}r_1 - \frac{\partial r_1}{\partial z_2} - \frac{\partial r_1}{\partial \eta}r_2\right]\frac{\partial}{\partial \eta},$$

which is not equal to 0 if $r_1 \neq 0$ and $\nu \neq 0$. However, we have $[\tau_1, \tau_2] \in \Delta^{\perp}$.

The following will find two other vector fields which are commutative, according to Lemma 7.4. For this, let us take $H_1 = -r_1\frac{\partial}{\partial \eta}$ and $H_2 = -r_2\frac{\partial}{\partial \eta}$, and then we obtain the desired result $\bar{\tau}_1 = \frac{\partial}{\partial z_1}$ and $\bar{\tau}_2 = \frac{\partial}{\partial z_2}$ which yields $[\bar{\tau}_1, \bar{\tau}_2] = 0$. □

Now as we did in (3.10) in Chap. 3, let us recall the procedure to compute the change of coordinates. For this, we consider r independent vector fields τ_i for $1 \leq i \leq r$, and define the following matrix:

$$\Lambda_1 = \begin{pmatrix} \theta_1 \\ \vdots \\ \theta_r \end{pmatrix} (\tau_1, \ldots, \tau_r) = \begin{pmatrix} 0 & 0 & \cdots & 0 & 1 \\ 0 & 0 & \cdots & 1 & \Lambda_{2,r} \\ \vdots & \vdots & \iddots & \vdots & \vdots \\ 0 & 1 & \cdots & \Lambda_{r-1,r-1} & \Lambda_{r-1,r} \\ 1 & \Lambda_{r,2} & \cdots & \Lambda_{r,r-1} & \Lambda_{r,r} \end{pmatrix}, \tag{7.13}$$

where its entries are $\Lambda_{i,j} = \theta_i(\tau_j)$ for $1 \leq i, j \leq r$. Due to the fact that this matrix is invertible, we can then define the following 1-forms:

$$\begin{pmatrix} \omega_1 \\ \vdots \\ \omega_r \end{pmatrix} = \Lambda_1^{-1}[\theta_1, \ldots, \theta_r]^T, \tag{7.14}$$

where its components are linear combination of the 1-forms θ_i for $1 \leq i \leq r$ and given by

$$\begin{aligned} \omega_r &= \theta_1 \\ \omega_{r-k} &= \left(\theta_{k+1} - \sum_{i=r-k+1}^{n} \Lambda_{k+1,i}\omega_i\right) \quad \text{for } 1 \leq k \leq r-1. \end{aligned} \tag{7.15}$$

From (7.14) and (7.15), we can deduce that 1-form ω_k for $1 \leq k \leq r$ are independent of the choice of τ_i modulo $\Delta^{\perp}$. From this, we can build the change of coordinates as state in the following theorem.

Theorem 7.1 *There exists a local change of coordinates $(z^T, \zeta^T)^T = \phi(x)$ on $\mathscr{X}$ which transforms system (7.1) into the target normal form (7.3), if and only if the following conditions are fulfilled*

(1) there exists a family of vector fields $\bar{\tau}_i = \tau_i$ modulo $\Delta^{\perp}$ such that $[\bar{\tau}_i, \bar{\tau}_l] = 0$ for $1 \leq i \leq r$ and $1 \leq l \leq r$;
(2) $\left[\bar{\tau}_i, \bar{\tau}_j\right] = 0$ for $1 \leq i \leq r$ and $r+1 \leq j \leq n$ where $\bar{\tau}_j$ was defined in C2 of Lemma 7.1.

Under the above conditions, the change of coordinates ϕ is locally determined by its differential $\phi_* := d\phi = \omega$, i.e., $\phi(x) = \int_{\gamma} \omega$, where γ is any path on $\mathscr{X}$ with $\gamma(0) = 0$ and $\gamma(1) = x$.

Proof **Necessity**: The proof is the same as those for the normal forms obtained in the previous chapters. Let us just specify the main difference with the normal form (7.3).

Let us denote the first partial observability 1-form for the normal form (7.3) by $\theta_1^{nf} = dz_r$ and the others have the following form

$$\theta_i^{nf} = dz_{r-i+1} + \sum_{j=1}^{i-1} l_j \theta_j^{nf}$$

for $2 \leq i \leq r$. Therefore the associated distribution is given by

$$\Delta^{nf\perp} = \text{span}\left\{\frac{\partial}{\partial \zeta_1}, \ldots, \frac{\partial}{\partial \zeta_{n-r}}\right\}.$$

Hence, the first vector field is $\tau_1^{nf} = \frac{\partial}{\partial z_1} + H_1^{nf}$ where $H_1^{nf} \in \Delta^{nf\perp}$, and by induction we build the rest vector fields as

$$\tau_i^{nf} = \frac{\partial}{\partial z_i} + H_i^{nf},$$

where $H_i^{nf} \in \Delta^{nf\perp}$ for $2 \leq i \leq r$. From this, we deduce the following family of vector fields

$$\bar{\tau}_i^{nf} = \frac{\partial}{\partial z_i} = \tau_i^{nf} \text{ modulo } \Delta^{nf\perp}$$

such that $\left[\bar{\tau}_i^{nf}, \bar{\tau}_l^{nf}\right] = 0$ for $1 \leq i \leq r$ and $1 \leq l \leq r$.

As we want them to commute, we can then choose $\bar{\tau}_j^{nf} = \frac{\partial}{\partial \zeta_{j-r}}$, thus we have $\left[\bar{\tau}_i^{nf}, \bar{\tau}_j^{nf}\right] = 0$ for $1 \leq i \leq r$ and $r+1 \leq j \leq n$. To complete the proof, it is enough to set $\omega = d\phi = \phi^* [dz, d\zeta]^T$, where ϕ^* is the pullback defined from ϕ on the space of 1-forms.

Sufficiency: Hereafter, we will state only the main difference with the full observability case stated in Chap. 3, and we will just give the key steps which enable us to adapt the proof of Theorem 3.2.

For this, we consider the following two cases:

- for $1 \leq i \leq r - 1$, we have

$$\frac{\partial}{\partial z_i}\phi_*(f) = \left[\frac{\partial}{\partial z_i}, \phi_*(f)\right] = [\phi_*(\bar{\tau}_i), \phi_*(f)] = \phi_*[\bar{\tau}_i, f] = \phi_*(\bar{\tau}_{i+1} + H_i)$$

for some $H_i \in \Delta^{\perp}$. Thus, we have

$$\frac{\partial}{\partial z_i}\phi_*(f) = \phi_*(\bar{\tau}_{i+1}) + \phi_*(H_i) = \frac{\partial}{\partial z_{i+1}} + \phi_*(H_i); \tag{7.16}$$

- for $r + 1 \leq j \leq n$, we have

$$\frac{\partial}{\partial \zeta_j}\phi_*(f) = \left[\frac{\partial}{\partial \zeta_j}, \phi_*(f)\right] = [\phi_*(\bar{\tau}_j), \phi_*(f)] = \phi_*[\bar{\tau}_j, f].$$

Since $\bar{\tau}_j$ span the distribution $\Delta^{\perp}$ which is f-invariant (see C3 of Lemma 7.1), then there exists $H_j \in \Delta^{\perp}$ such that $H_j = [\bar{\tau}_j, f]$. Thus, we have

$$\frac{\partial}{\partial \zeta_j}\phi_*(f) = \phi_*(H_j). \tag{7.17}$$

As for any $H \in \Delta^{\perp}$, we know that

$$\phi_*(H) = \omega(H) \in \text{span}\left\{\frac{\partial}{\partial \zeta_1}, \ldots, \frac{\partial}{\partial \zeta_{n-r}}\right\}.$$

The remainder of the proof can be obtained in the same way as that of Theorem 3.2 in Chap. 3. □

Let us present an example to show how we deal with those conditions stated in Theorem 7.1.

Example 7.2 Consider the following nonlinear system:

$$\begin{aligned} \dot{x}_1 &= -x_3^2 + x_1^3 x_3 - \tfrac{1}{2}x_3^3 + x_2^5 \\ \dot{x}_2 &= x_1 - \tfrac{1}{2}x_3^2 \\ \dot{x}_3 &= -x_3 + x_1^3 - \tfrac{1}{2}x_3^2 \\ y &= x_2. \end{aligned} \tag{7.18}$$

A simple calculation leads to the following partial observability 1-forms:

$$\begin{aligned}\theta_1 &= dx_2\\ \theta_2 &= dx_1 - x_3 dx_3.\end{aligned}$$

Therefore, we have

$$\text{rank}\left\{dh, dL_f h, dL_f^2 h\right\} = 2,$$

thus $r = 2$ because $dL_f^2 h = 5x_2^4\theta_1$.

From this, we have the co-distribution $\Delta = \text{span}\,\{\theta_1, \theta_2\}$, and the associated distribution

$$\Delta^\perp = \text{span}\left\{x_3\frac{\partial}{\partial x_1} + \frac{\partial}{\partial x_3}\right\}.$$

Due to the fact that $\bar{\tau}_3$ needs to be a commutative basis of $\Delta^\perp$, we can then choose

$$\bar{\tau}_3 = x_3\frac{\partial}{\partial x_1} + \frac{\partial}{\partial x_3}.$$

As a solution of (7.7), we obtain

$$\tau_1 = \frac{\partial}{\partial x_1} + r_1(x)\bar{\tau}_3.$$

By removing all terms in the direction of $\bar{\tau}_3$ in τ_1, we get

$$\bar{\tau}_1 = \frac{\partial}{\partial x_1}.$$

Now, we compute

$$[\bar{\tau}_1, f] = \frac{\partial}{\partial x_2} + 3x_1^2\bar{\tau}_3,$$

and we cancel all terms in the direction of $\bar{\tau}_3$ to obtain

$$\bar{\tau}_2 = \frac{\partial}{\partial x_2}.$$

As a result, we obtain the following vector fields

$$\begin{aligned}\bar{\tau}_1 &= \tfrac{\partial}{\partial x_1}\\ \bar{\tau}_2 &= \tfrac{\partial}{\partial x_2}\\ \bar{\tau}_3 &= x_3\tfrac{\partial}{\partial x_1} + \tfrac{\partial}{\partial x_3}\end{aligned}$$

that fulfill all conditions of Theorem 7.1, therefore there exits a change of coordinates which can transform the studied system into the target form (7.3).

For the deduction of this diffeomorphism, we use the obtained family of vector fields $\bar{\tau} = [\bar{\tau}_1, \bar{\tau}_2, \bar{\tau}_3]$ which gives

$$\Lambda_1 = \begin{bmatrix} \theta_1 \\ \theta_2 \end{bmatrix} (\bar{\tau}_1, \bar{\tau}_2) = \begin{pmatrix} 0 & 1 \\ 1 & 0 \end{pmatrix}.$$

Thus, we have

$$\begin{pmatrix} \omega_1 \\ \omega_2 \end{pmatrix} = \Lambda_1^{-1} \begin{pmatrix} \theta_1 \\ \theta_2 \end{pmatrix} = \begin{pmatrix} dz_1 \\ dz_2 \end{pmatrix} = d \begin{pmatrix} x_1 - \frac{1}{2}x_3^2 \\ x_2 \end{pmatrix}.$$

Therefore, we get the following change of coordinates:

$$\phi(x) = \begin{pmatrix} x_1 - \frac{1}{2}x_3^2 \\ x_2 \\ x_3 \end{pmatrix},$$

which transforms (7.18) into the following normal form:

$$\begin{aligned} \dot{z}_1 &= 0 \\ \dot{z}_2 &= z_1 \\ \dot{\zeta}_1 &= z_1^3 - z_3 + \tfrac{z_3^2}{2} \left(3z_1^2 + \tfrac{3}{2}z_1 z_3^2 + \tfrac{1}{4}z_3^4 - 1\right) \\ y &= z_2. \end{aligned}$$

□

7.3 Diffeomorphism on the Output

Let us remark that the deduced change of coordinates $\phi(x)$ in Sect. 7.2 does not modify the output. This is due to the fact that $\theta_r(\tau_1) = 1$. In Chap. 4, we have seen that if the commutativity conditions of Lie brackets are not fulfilled, one way to relax this constraint is to seek a new vector field, noted as $\bar{\sigma}_1$, such that $\theta_r(\bar{\sigma}_1)$ becomes a function of the output, and this will introduce a diffeomorphism on the output.

For this purpose, we will follow the same algorithm as in Chap. 4. Indeed will adapt the normal form (4.3) obtained in Chap. 4 together with the normal form (7.3) as follows:

$$\begin{aligned} \dot{z} &= A_O z + \beta(\bar{y}) \\ \dot{\zeta} &= \eta(z, \zeta) \\ \bar{y} &= C_O z = \varphi(y). \end{aligned} \tag{7.19}$$

As in Sect. 7.2, we calculate the observability 1-forms θ_i for $1 \le i \le r$ where r is the rank of observability matrix, which defines the co-distribution

$$\Delta = \text{span}\{\theta_1, \theta_2, \ldots, \theta_r\}$$

and the kernel (or the annihilator) of the co-distribution Δ, noted as $\Delta^{\perp}$. We assume that we have already obtained a family of vector fields $\bar{\tau}_j$ for $r+1 \leq j \leq n$ which are a commutative basis of $\Delta^{\perp}$, and $\bar{\tau}_i$ for $1 \leq i \leq r$ without any component in the direction of $\bar{\tau}_j$ for $r+1 \leq j \leq n$. Thus, hereafter, we assume that there exist $1 \leq i, l \leq r$ such that $[\bar{\tau}_i, \bar{\tau}_l] \neq 0$, hence Theorem 7.1 cannot be applied. As it was explained in Chap. 4, in order to introduce a diffeomorphism on the output, let $l(y) \neq 0$ be a function of the output which will be determined later. Then we can define a new vector field σ_1 from $\bar{\tau}_1$ as follows:

$$\bar{\sigma}_1 = l(y)\bar{\tau}_1. \tag{7.20}$$

Like what we did in the previous chapters, we define by induction the following vector fields:

$$\bar{\sigma}_i = \left[\bar{\sigma}_{i-1}, f\right] \text{ modulo } \Delta^{\perp} \tag{7.21}$$

for $2 \leq i \leq r$. As in the proof of Lemma 4.1 of Chap. 4, a straightforward calculation leads to

$$\bar{\sigma}_i = \sum_{k=1}^{i} (-1)^{i-k} C_{i-1}^{k-1} L_f^{i-k} l(y)\bar{\tau}_k \text{ modulo } \Delta^{\perp}, \tag{7.22}$$

where C_{i-1}^{k-1} are the binomial coefficients and $L_f^0 l(y) = l(y)$.

By canceling all terms in the directions of $\bar{\tau}_j$ for $r+1 \leq j \leq n$ to get the commutative basis of $\Delta^{\perp}$, we obtain

$$\sigma_i = \sum_{k=1}^{i} (-1)^{i-k} C_{i-1}^{k-1} L_f^{i-k} l(y)\bar{\tau}_k, \text{ for } 1 \leq i \leq r. \tag{7.23}$$

Now, let us set

$$\sigma_j = \bar{\tau}_j, \text{ for } r+1 \leq j \leq n, \tag{7.24}$$

where $\{\bar{\tau}_{r+1}, \ldots, \bar{\tau}_n\}$ are commutative basis of $\Delta^{\perp}$. Then, as in Chap. 4, let us define the following matrix:

$$\tilde{\Lambda}_1 = \begin{pmatrix} \theta_1 \\ \vdots \\ \theta_r \end{pmatrix} (\sigma_1, \ldots, \sigma_r) = \begin{pmatrix} 0 & \cdots & 0 & l(y) \\ \vdots & \cdots & l(y) & * \\ 0 & \cdots & * & * \\ l(y) & \cdots & * & * \end{pmatrix}, \tag{7.25}$$

which is invertible because $l(y) \neq 0$, and therefore we can define the following 1-forms:

$$\begin{pmatrix} \tilde{\omega}_1 \\ \vdots \\ \tilde{\omega}_r \end{pmatrix} = \tilde{\Lambda}_1^{-1} [\theta_1, \ldots, \theta_r]^T \tag{7.26}$$

and using the same method in Sect. 7.2, we can uniquely determine $\tilde{\omega}_i$ for $r+1 \leq i \leq n$ by solving the following equations:

$$\tilde{\omega}_i(\sigma_j) = \delta_i^j \tag{7.27}$$

for $1 \leq j \leq n$. Finally, we have the following theorem.

Theorem 7.2 *There exists a local change of coordinates $(z^T, \zeta^T)^T = \phi(x)$ on $\mathscr{X}$ which transforms system (7.1) into the target normal form (7.19) if and only if the following conditions are fulfilled:*

(1) $[\sigma_i, \sigma_l] = 0$ for $1 \leq i \leq r$ and $1 \leq l \leq r$ where σ_i defined in (7.23);
(2) $\left[\sigma_i, \sigma_j\right] = 0$ for $1 \leq i \leq r$ and $r+1 \leq j \leq n$ where σ_j defined in (7.24).

Under the above conditions, the change of coordinates ϕ is locally given by its differential $\phi_* := d\phi = \tilde{\omega}$, i.e.,

$$\phi(x) = \int_\gamma \tilde{\omega}, \tag{7.28}$$

where γ is any path on $\mathscr{X}$ with $\gamma(0) = 0$ and $\gamma(1) = x$. Moreover, we have $z_r = \bar{y} = \varphi(y)$ where $\varphi(y) = \int_0^y \frac{1}{l(c)} dc$ is a change of coordinates on the output.

Proof The proof of Theorem 7.2 is similar to that of Theorem 7.1, thus the following just gives hints to determine $l(y)$ and some different elements with respect to the result presented in Chap. 4.

According to (7.20), we define $\sigma_1 = l(y)\bar{\tau}_1$ where $\bar{\tau}_1$ is obtained in Sect. 7.2 by canceling all terms in the directions of $\bar{\tau}_j$ for $r+1 \leq j \leq n$. We obtain by induction

$$\sigma_i = \left[\sigma_{i-1}, f\right] \text{ modulo } \Delta^\perp$$

for $2 \leq i \leq r$. Then we can deduce

$$\begin{aligned}
[\sigma_1, \sigma_i] &= \sum_{k=1}^{i} (-1)^{i-k} C_{i-1}^{k-1} L_f^{i-k} l(y) [\sigma_1, \bar{\tau}_k] \\
&= \sum_{k=1}^{i} (-1)^{i-k} C_{i-1}^{k-1} L_f^{i-k} l(y) \left[l(y) [\bar{\tau}_1, \bar{\tau}_k] - L_{\bar{\tau}_k} l(y) \bar{\tau}_1\right] \\
&\quad + \sum_{k=1}^{i} (-1)^{i-k} C_{i-1}^{k-1} L_{\sigma_1} L_f^{i-k} l(y) \bar{\tau}_k .
\end{aligned} \tag{7.29}$$

Now, we compute $l(y)$, by imposing $[\sigma_1, \sigma_i] = 0$. More details on this issue can be founded in Chap. 4. After having deduced $l(y)$, the rest of the proof for Theorem 7.2 is similar to that of Theorem 7.1. □

7.4 Observer Design

7.4.1 Homogeneous Observer

In this section, we will present a homogeneous non-asymptotic observer for (7.1). The global finite-time observer based on homogeneity [1, 2] was firstly introduced by [6] for the nonlinear dynamical systems which can be transformed into a linear dynamical system with output injection. After that, [8] extended this idea and proposed a semi-global finite-time observer for the special systems with triangular structure. The global finite-time observer for such a system was studied in [4] by introducing the second gain in [9].

Let us suppose that there exists a diffeomorphism $(z^T, \zeta^T)^T = \phi(x)$ on $\mathscr{X}$ which transforms system (7.1) into the partial observer normal form (7.19). To design non-asymptotic observer of z in (7.19), we can consider the following dynamics:

$$\dot{\hat{z}} = A_O\hat{z} + \beta(\bar{y}) + K\lceil\varphi(y) - C_O\hat{z}\rfloor^{\alpha}, \tag{7.30}$$

where $K = [K_1, \dots, K_n]$ is a constant matrix to be chosen such that the matrix $(A_O - KC_O)$ is Hurwitz, and $\lceil\varphi(y) - C_O\hat{z}\rfloor^{\alpha}$ is defined as follows:

$$\lceil\varphi(y) - C_O\hat{z}\rfloor^{\alpha} = \begin{bmatrix} |\varphi(y) - C_O\hat{z}|^{\alpha_1}\text{sgn}(\varphi(y) - C_O\hat{z}) \\ \vdots \\ |\varphi(y) - C_O\hat{z}|^{\alpha_n}\text{sgn}(\varphi(y) - C_O\hat{z}) \end{bmatrix}, \tag{7.31}$$

where α_j for $1 \le j \le n$ are defined as

$$\alpha_j = j\gamma - (j-1) \tag{7.32}$$

with $\gamma \in (1 - \frac{1}{n}, 1)$.

Note $e = z - \hat{z}$, then according to (7.19) and (7.30), we obtain

$$\begin{bmatrix} \dot{e}_1 \\ \dot{e}_2 \\ \vdots \\ \dot{e}_n \end{bmatrix} = \begin{bmatrix} 0 & 0 & \cdots & 0 & 0 \\ 1 & 0 & \cdots & 0 & 0 \\ \vdots & \vdots & \vdots & \ddots & \vdots \\ 0 & 0 & \cdots & 1 & 0 \end{bmatrix} \begin{bmatrix} e_1 \\ e_2 \\ \vdots \\ e_n \end{bmatrix} - \begin{bmatrix} K_1|e_n|^{\alpha_1}\text{sgn}(e_n) \\ K_2|e_n|^{\alpha_2}\text{sgn}(e_n) \\ \vdots \\ K_n|e_n|^{\alpha_n}\text{sgn}(e_n) \end{bmatrix}$$

and it has been proven in [6] that e will converge to zero in a finite time $T(z(0))$, i.e.,

$$\hat{z}(t) = z(t), \forall t \ge T(z(0)).$$

In conclusion, the dynamics (7.30) is a non-asymptotic homogeneous observer of (7.19).

7.4.2 Design Procedure

The following summarizes the procedure to design observer based on the proposed normal form (7.19).

Step 1: Determine the observability 1-forms θ_i for $1 \leq i \leq r$;
Step 2: Pick a solution τ_1 of (7.7), then cancel all its terms in the directions of $\{\bar{\tau}_{r+1}, \ldots, \bar{\tau}_n\}$ which yields $\bar{\tau}_1$;
Step 3: Iteratively, for $2 \leq i \leq r$, compute $[\bar{\tau}_{i-1}, f]$ in which eliminate all terms in the directions of $\{\bar{\tau}_{r+1}, \ldots, \bar{\tau}_n\}$ which gives $\bar{\tau}_i$;
Step 4: Calculate the new family of vector fields $\bar{\sigma}$ according to (7.20) and (7.21);
Step 5: Deduce $l(y)$ according to (7.29);
Step 6: Compute $\tilde{\Lambda}$ and $\tilde{\omega}$ according to (7.25) and (7.26), then deduce the change of coordinates $\phi(x)$ according to (7.28);
Step 7: Apply this change of coordinates to calculate $\phi_*(f)$, with f being defined in (7.1);
Step 8: Design the proposed homogeneous observer (7.30) to estimate z of (7.19).

References

1. Bhat, S., Bernstein, D.: Finite-time stability of continuous autonomous systems. SIAM J. Control Optim. **38**(3), 751–766 (2000)
2. Bhat, S., Bernstein, D.: Geometric homogeneity with applications to finite-time stability. Math. Control Signals Syst. **17**(2), 101–127 (2005)
3. Jo, N.H., Seo, J.H.: Observer design for non-linear systems that are not uniformly observable. Int. J. Control **75**(5), 369–380 (2002)
4. Menard, T., Moulay, E., Perruquetti, W.: A global high-gain finite-time observer. IEEE Trans. Autom. Control **55**(6), 1500–1506 (2010)
5. Nijmeijer, H., Van der Schaft, A.: Nonlinear Dynamical Control Systems, vol. 175. Springer, Berlin (1990)
6. Perruquetti, W., Floquet, T., Moulay, E.: Finite-time observers: application to secure communication. IEEE Trans. Autom. Control **53**(1), 356–360 (2008)
7. Röbenack, K., Lynch, A.: Observer design using a partial nonlinear observer canonical form. Int. J. Appl. Math. Comput. Sci. **16**, 333–343 (2006)
8. Shen, Y., Xia, X.: Semi-global finite-time observers for nonlinear systems. Automatica **44**(12), 3152–3156 (2008)
9. Shen, Y., Huang, Y., Gu, J.J.: Global finite-time observers for Lipschitz nonlinear systems. IEEE Trans. Autom. Control **56**(2), 418–424 (2011)
10. Tami, R., Zheng, G., Boutat, D., Aubry, D.: Partial observability normal form for nonlinear functional observers design. In: Proceedings of the 32nd IEEE Chinese Control Conference, pp. 95–100 (2013)
11. Tami, R., Zheng, G., Boutat, D., Aubry, D., Wang, H.: Partial observer normal form for nonlinear system. Automatica **64**, 54–62 (2016)
12. Trinh, H., Fernando, T., Nahavandi, S.: Partial-state observers for nonlinear systems. IEEE Trans. Autom. Control **51**(11), 1808–1812 (2006)

Chapter 8
Extension to Nonlinear Dynamical Systems with Multiple Outputs

Abstract Till now, all the observer normal forms presented in Chaps. 3–7 are focused on nonlinear dynamical systems with single output. The extension of the previous results to the case with multiple outputs is not trivial [3, 13]. The first attempt was made by [6, 10], and this chapter will firstly summarize their ideas, and then discuss the main differences between single output and multiple outputs. After that, some special cases related to observability indices will be discussed for nonlinear dynamical systems with multiple outputs, for which necessary and sufficient geometric conditions will be presented. Another extension treated in this chapter is the partially observable case, which has been analyzed in Chap. 7 for systems with single output. By following a similar procedure as that in Chap. 7, a desired partial observer normal form with multiple outputs is then proposed [8]. Of course, we give as well the geometric conditions which enable us to compute a diffeomorphism to transform a nonlinear system with multiple outputs into such a normal form, for which a reduced-order Luenberger-like observer is proposed to estimate its observable states.

8.1 Problem Statement

Consider the following multi-output nonlinear dynamical system

$$\begin{aligned} \dot{x} &= f(x) \\ y &= h(x) = (h_1(x), \ldots, h_m(x))^T, \end{aligned} \tag{8.1}$$

where $x \in \mathscr{X} \subseteq \mathbb{R}^n$ is the state vector, $\mathscr{X}$ is an open domain of $\mathbb{R}^n$, $f : \mathscr{X} \to \mathbb{R}^n$ and $h : \mathscr{X} \to \mathbb{R}^m$ are both assumed to be smooth.

Without loss of generality, it is assumed that $0 \in \mathscr{X}$, $f(0) = 0$ and $h(0) = 0$. Moreover, it is assumed that all components of h: $y_i = h_i$ for $1 \le i \le m$, are independent. Otherwise, we can just drop the dependent components and keep only the maximal independent components.

For the case with multiple outputs, this chapter investigates firstly the problem how to seek a change of coordinates $z = \phi(x)$ with $z = (z_1^T, z_2^T, \ldots, z_m^T)^T$ and

D. Boutat and G. Zheng, *Observer Design for Nonlinear Dynamical Systems*, Lecture Notes in Control and Information Sciences 487,
https://doi.org/10.1007/978-3-030-73742-9_8

$z_i = \left(z_{i,1}, z_{i,2}, \ldots, z_{i,r_i}\right)^T$, for $1 \leq i \leq m$ with $\sum_{i=1}^{m} r_i = n$, which can transform the dynamical system (8.1) into the following observer normal form:

$$\begin{aligned} \dot{z}_i &= A_{O_i} z_i + \beta_i(\bar{y}) \\ \bar{y}_i &= C_{O_i} z_i \end{aligned} \tag{8.2}$$

for $1 \leq i \leq m$, where A_{O_i} and C_{O_i} are $(r_i \times r_i)$ and $(1 \times r_i)$ constant matrices, respectively, of the following form:

$$A_{O_i} = \begin{bmatrix} 0 & \cdots & 0 & 0 \\ 1 & \cdots & 0 & 0 \\ \vdots & \ddots & \vdots & \vdots \\ 0 & \cdots & 1 & 0 \end{bmatrix}, \quad C_{O_i} = \left(0, \ldots, 0, 1\right), \tag{8.3}$$

i.e., the pairs (A_{O_i}, C_{O_i}) are of the well-known Brunovsky form. Moreover, $\bar{y} = (\bar{y}_1, \ldots, \bar{y}_m) = \varphi(y)$, where φ is a local diffeomorphism of the original outputs, and $\beta_i = \left(\beta_{i,1}, \beta_{i,2}, \ldots, \beta_{i,r_i}\right)^T$, for $1 \leq i \leq m$, are vectors of functions of the new outputs $\bar{y}$.

8.2 Construction of the Frame

Note that in Chap. 3, for the single-output case, we had constructed a family of vector fields τ in (3.4) and (3.5) for the purpose of producing the desired change of coordinates. For the multiple outputs case studied in this chapter, we will follow the same procedure.

Unlike the single-output case where the observability rank condition depends only on the single output and its high-order derivatives, it becomes more complicated for the multiple-output case since the high-order derivatives of some outputs might be dependent. To characterize this dependence, the so-called observability indices have been introduced in the literature when analyzing the dynamical systems with multiple outputs.

For this, denote

$$D_0 = \text{span}\left\{dh_i, \ \text{for } 1 \leq i \leq m\right\},$$

which is spanned by the differential of the outputs, and it satisfies

$$\text{rank} D_0 = m.$$

Set $m_0 = m$, then for $k \geq 1$, define the following family of co-distributions

$$D_k = \text{span}\left\{dL_f^{j-1} h_i, \ \text{for } 1 \leq i \leq m, 1 \leq j \leq k+1\right\}.$$

Hence, for $j \geq 1$, we have $D_{j-1} \subseteq D_j$, and denote the difference of their ranks as

$$m_j = \text{rank} D_j - \text{rank} D_{j-1}.$$

Using these notations, the observability indices can be then defined as follows.

Definition 8.1 ([6]) The observability indices of the pair (f, h) are defined as

$$r_i := \text{card}\{m_j | \, i \leq m_j, \; 0 \leq j \leq m\} \tag{8.4}$$

for $1 \leq i \leq m$. With a possible reordering of the output indices, it can be assumed that $r_1 \geq r_2 \geq \cdots \geq r_m$.

Based on the concept of observability indices, the following assumption is imposed for the studied system (8.1).

Assumption 8.1 For the pair (f, h) defined in system (8.1), it is supposed that the observability rank condition is satisfied in the sense that the rank of the following co-distribution

$$\Delta = \text{span}\left\{dL_f^{k-1} h_i, \; 1 \leq i \leq m, \; 1 \leq k \leq r_i\right\} \tag{8.5}$$

is equal to n, i.e., $\sum_{i=1}^{m} r_i = n$. Without loss of generality, assume as well that the observability indices $\{r_i\}_{1 \leq i \leq m}$ of (8.1) are constant on $\mathscr{X}$.

Like in the previous chapters, we define $\theta_{i,k} = dL_f^{k-1} h_i$, then we can construct the following multi-valued 1-form:

$$\theta = (\theta_{i,k})_{1 \leq i \leq m, 1 \leq k \leq r_i} = \begin{pmatrix} \theta_{1,1} \\ \vdots \\ \theta_{1,r_1} \\ \theta_{2,1} \\ \vdots \\ \theta_{m,r_m} \end{pmatrix}. \tag{8.6}$$

Now, let us define a family of vector fields $\left\{\tau_{i,1}\right\}_{1 \leq i \leq m}$ by solving the following algebraic equations:

$$\begin{array}{ll} \theta_{j,k}\left(\tau_{i,1}\right) = 0, \; \text{for } 1 \leq k \leq r_i - 1, & \text{if } j \leq i - 1 \\ \begin{cases} \theta_{j,r_j}\left(\tau_{i,1}\right) = 1 \\ \theta_{j,k}\left(\tau_{i,1}\right) = 0, \; \text{for } 1 \leq k \leq r_j - 1 \end{cases} & \text{if } j = i, \\ \theta_{j,k}\left(\tau_{i,1}\right) = 0, \; \text{for } 1 \leq k \leq r_j, & \text{if } j \geq i + 1 \end{array} \tag{8.7}$$

for $1 \leq i \leq m$ and $1 \leq j \leq m$.

With the above solution, by applying the Lie bracket $[\tau_{i,k-1}, f]$ between vector fields $\tau_{i,k-1}$ and f, we can define the following vector fields:

$$\tau_{i,k} = [\tau_{i,k-1}, f] \tag{8.8}$$

for $1 \le i \le m$ and $2 \le k \le r_i$.

Since it is assumed that Assumption 8.1 is fulfilled, then the family of vector fields $\{\tau_{i,1}\}_{1\le i\le m}$ defined in (8.7) and $\tau_{i,k}$ defined in (8.8) yield the following frame

$$\tau = \left(\tau_{i,j}\right)_{1\le i\le m,\ 1\le j\le r_i} = (\tau_{1,1}, \ldots, \tau_{1,r_1}, \ldots, \tau_{m,1}, \ldots, \tau_{m,r_m})$$

of the tangent fiber bundle $T\mathscr{X}$ of $\mathscr{X}$. It is clear that, for $1 \le i \le m$ and $2 \le k \le r_i$, we always have

$$\theta_{j,k}\left(\tau_{i,s}\right) = 0, \text{ if } r_k > r_s.$$

We would like emphasize that the solution of (8.7) is generally not unique. Multiple solutions might exist even for the case when the dynamical system is fully observable, i.e., $\sum_{i=1}^{m} r_i = r = n$. In fact, the first vector field (i.e., $\tau_{1,1}$) should satisfy r equations (for the fully observable case with $r = n$, $\tau_{1,1}$ is then unique), and $\tau_{i,1}$ for $2 < i < m$ should satisfy only $ir_i + \sum_{k=i+1}^{m} r_k = r - \sum_{j<i}^{m}(r_j - r_i) + \sum_{k=i+1}^{m} r_k$ equations. So the number of algebraic equations is less than that of the variables to be solved, thus the solutions are not unique.

8.3 Necessary and Sufficient Conditions

Consider now the following co-distributions

$$\pounds_i = \left\{dL_f^{k-1}h_j, \text{ for } 1 \le j \le m \text{ and } 1 \le k \le r_i\right\} \setminus \{dL_f^{r_i-1}h_i\} \tag{8.9}$$

for $1 \le i \le m$, where the symbol $\setminus$ stands for the exclusion. Let us firstly consider the fully observable case (the partial observability case will be analyzed in Sect. 8.6), i.e., Assumption 8.1 is fulfilled for the studied dynamical system (8.1). Thus, Δ defined in (8.5) is of full rank, i.e.,

$$\dim\Delta = n.$$

Given a nonlinear dynamical system with multiple outputs, when it is fully observable, then the following result has been stated in [6, 10].

Theorem 8.1 *There exists a local diffeomorphism $z = \phi(x)$ via which system (8.1) can be transformed into (8.2) if and only if*

(1) there exists a frame $\tau_{j,s}$ given by Eqs. (8.7)–(8.8) such that

$$\left[\tau_{i,k}, \tau_{j,s}\right] = 0 \tag{8.10}$$

for all $1 \le i,\ j \le m,\ 1 \le k \le r_i$ *and* $1 \le s \le r_j$*;*

(2) for $1 \le i \le m$ *we have*

$$span\left\{£_i\right\} = span\left\{£_i \cap £\right\}, \tag{8.11}$$

where Δ *is defined in (8.5) and* $£_i$ *for* $1 \le i \le m$ *are defined in (8.9).*

Let us clarify the meaning of (8.11) in Theorem 8.1, which indeed is to ensure that the outputs will not be influenced by their derivatives. Precisely, suppose that there exists $1 \le i \le m$ such that

$$\mathrm{span}\left\{£_i\right\} \ne \mathrm{span}\left\{£_i \cap \Delta\right\},$$

then we can find $j > i$ with $r_i > r_j$ and $1 \le k \le r_i - r_j$ such that

$$dL_f^{r_j-1+k} h_j \notin \mathrm{span}\left\{£_i \cap £\right\}. \tag{8.12}$$

Due to the fact the studied system is fully observable, we get

$$dL_f^{r_j-1+k} h_j = \sum_{s=1}^{i-1}\left(\sum_{t=1, r_s > r_i}^{r_s - r_i} \lambda_{s,r_i+t}\theta_{s,r_i+t}\right) + \lambda_{i,r_i}\theta_{i,r_i} + \mu, \tag{8.13}$$

where $\mu = \sum_{s=1}^{i}\sum_{t=1}^{r_i} \beta_{s,t}\theta_{s,t}$ with $\beta_{i,r_i} = 0$, which contains all the terms with the indices satisfying $r_s \le r_i$.

It is obvious that

$$\mu \in \mathrm{span}\left\{£_i \cap £\right\}$$

and

$$\theta_{s,r_i+t} \in £ \setminus £_i \cap £.$$

Particularly we obtain

$$\mu(\bar{\tau}_{i,1}) = 0$$

and

$$\mu(\bar{\tau}_{s,r_s-(r_i+t)+1}) = 0$$

for $1 \le s \le i-1$ and $0 \le t \le r_s - r_i$.

Concerning Eq. (8.12), the following result presents an alternative to verify it.

Proposition 8.1 *For* $1 \le i < j \le m$*, the following three statements are equivalent:*

(1) for $1 \le k \le r_i - r_j$*,*

$$dL_f^{r_j-1+k}h_j \in span\,\{£_i \cap £\};$$

(2) *for* $1 \leq s \leq i-1$, $0 \leq t \leq r_s - r_i$ *and* $1 \leq k \leq r_i - r_j$,

$$\begin{cases} dL_f^{r_j-1+k}h_j(\bar{\tau}_{i,1}) = 0 \\ dL_f^{r_j-1+k}h_j(\bar{\tau}_{s,r_s-(r_i+t)+1}) = 0; \end{cases}$$

(3) *for* $1 \leq s \leq i-1$, $0 \leq t \leq r_s - r_i$ *and* $1 \leq k \leq r_i - r_j$,

$$\begin{cases} dh_j(\bar{\tau}_{i,r_j+k}) = 0 \\ dh_j(\bar{\tau}_{s,r_s-(r_i+t)+r_j+k}) = 0. \end{cases}$$

Proof For $\tau_{j,s}$ defined in (8.7)–(8.8) and $£_i$ defined in (8.9), we can see that the distribution determined via the kernel of the co-distribution $span\,\{£_i \cap £\}$ is spanned by $\bar{\tau}_{i,1}$ and $\bar{\tau}_{s,r_s-(r_i+t)+1}$ for $1 \leq s \leq i-1$ and $0 \leq t \leq r_s - r_i$. Hence, according to Eq. (8.13), it is easy to conclude that the first statement of Proposition 8.1 is equivalent to the second one.

For the purpose of proving the equivalence between the second statement of Proposition 8.1 and the third one, we should firstly show that if $dL_f^l h_j(\bar{\tau}_{p,q}) = 0$, then

$$dL_f^{l+1}h_j(\bar{\tau}_{p,q}) = -dL_f^l h_j(\bar{\tau}_{p,q+1}). \tag{8.14}$$

In order to prove this, let us recall a basic result stated in Chap. 2, i.e., for a differential 1-form ν, its differential $d\nu$ is a differential 2-form, defined by its evaluation on a pair of vector fields X and Y as follows:

$$d\nu(X,Y) = L_X\nu(Y) - L_Y\nu(X) - \nu([X,Y]). \tag{8.15}$$

In (8.15), if we choose $\nu = dL_f^l h_j$, $X = f$ and $Y = \bar{\tau}_{p,q}$, due to the fact that $d\nu = 0$, i.e., an exact 1-form, then (8.15) can be rewritten as

$$0 = L_f dL_f^l h_j(\bar{\tau}_{p,q}) - L_{\bar{\tau}_{p,q}} dL_f^l h_j(f) - dL_f^l h_j([f, \bar{\tau}_{p,q}]).$$

Clearly, we can see that if $dL_f^l h_j(\bar{\tau}_{p,q}) = 0$, then

$$0 = -dL_f^{l+1}h_j(\bar{\tau}_{p,q}) + dL_f^l h_j(\bar{\tau}_{p,q+1})$$

and this is exactly the formulae (8.14) which we want to prove.

Since

$$dL_f^l h_j(\bar{\tau}_{p,q}) = 0 \text{ for } 1 \leq l \leq r_j - 2 + k$$

therefore, by induction we get

$$dL_f^{r_j-1+k}(\bar{\tau}_{p,q}) = dh_j(\bar{\tau}_{p,q+r_j+k-1}).$$

Finally, since $dL^l_f h_j(\bar{\tau}_{p,q}) = 0$ for $0 \leq l \leq r_j - 2 + k$, if we choose $\bar{\tau}_{p,q} = \bar{\tau}_{i,1}$ or $\bar{\tau}_{s,r_s-(r_i+t)+1}$, then we can prove that the second statement of Proposition 8.1 is equivalent to the third one.

We would like to remark that the third statement of Proposition 8.1 is very useful in practice due to the fact that it only contains the terms which have already been calculated (i.e., dh_j, $\bar{\tau}_{i,r_j+k}$ and $\bar{\tau}_{s,r_s-(r_i+t)+r_j+k}$). In other words, it is not necessary to compute high-order derivatives $dL_f^{r_j-1+k} h_j$ for $k \geq 1$. Besides, this shows that (8.11) in Theorem 8.1 is fundamental for the case where the outputs are not function of their derivatives in the proposed observer normal form (8.2).

The following example is to highlight the above discussion.

Example 8.1 Consider the following dynamical system:

$$
\begin{aligned}
\dot{x}_1 &= 0, \dot{x}_2 = x_1 \\
\dot{x}_3 &= x_2 + x_3 x_4 \\
\dot{x}_4 &= x_3, \dot{x}_5 = x_5 x_6 \\
\dot{x}_6 &= x_5 \\
y_1 &= x_4, \;\; y_2 = x_3 + x_6.
\end{aligned}
$$

Let us firstly calculate the corresponding observability 1-forms

$$
\begin{aligned}
\theta_{1,1} &= dx_4, \theta_{1,2} = dx_3 \\
\theta_{1,3} &= dx_2 + x_4 dx_3 + x_3 dx_4 \\
\theta_{1,4} &= dx_1 + d\left(x_4 x_2 + x_3 x_4^2 + x_3^2\right) \\
\theta_{2,1} &= dx_3 + dx_6 \\
\theta_{2,2} &= dx_5 + dx_2 + x_4 dx_3 + x_3 dx_4.
\end{aligned}
$$

By following Eqs. (8.7)–(8.8) and (8.9), we can have

$$
\begin{aligned}
\bar{\tau}_{1,1} &= \frac{\partial}{\partial x_1}, \bar{\tau}_{1,2} = \frac{\partial}{\partial x_2} \\
\bar{\tau}_{1,3} &= \frac{\partial}{\partial x_3}, \bar{\tau}_{1,4} = \frac{\partial}{\partial x_4} + x_4 \frac{\partial}{\partial x_3} \\
\bar{\tau}_{2,1} &= \frac{\partial}{\partial x_5}, \bar{\tau}_{2,2} = \frac{\partial}{\partial x_6} + x_6 \frac{\partial}{\partial x_5}
\end{aligned}
$$

and

$$
\begin{aligned}
£_1 &= \left\{\theta_{1,1}, \theta_{1,2}, \theta_{1,3}, \theta_{2,1}, \theta_{2,2}, dL_f^2 h_2, dL_f^3 h_2\right\} \\
£ &= \left\{\theta_{1,1}, \theta_{1,2}, \theta_{1,3}, \theta_{1,4}, \theta_{2,1}, \theta_{2,2}\right\} \\
£_1 \cap £ &= \left\{\theta_{1,1}, \theta_{1,2}, \theta_{1,3}, \theta_{2,1}, \theta_{2,2}\right\}.
\end{aligned}
$$

For the obtained vector fields $\bar{\tau}_{i,j}$, we can check that they commute. Therefore (8.10) of Theorem 8.1 is satisfied. Consequently, we can find the following diffeomorphism:

$$\begin{aligned} z_{1,1} &= x_1 \\ z_{1,2} &= x_2 \\ z_{1,3} &= x_3 - \tfrac{1}{2}x_4^2 \\ z_{1,4} &= x_4. \end{aligned}$$

Nevertheless, condition (8.11) of Theorem 8.1 is violated, since a straightforward computation yields

$$\begin{aligned} dL_f^2 h_2 &= \theta_{1,4} + x_6\left(\theta_{2,2} - \theta_{1,3}\right) + x_5\left(\theta_{2,1} - \theta_{1,3}\right) \\ &\notin \pounds_1 - (\pounds_1 \cap \pounds). \end{aligned}$$

In fact, we get

$$z_{2,1} = x_5 - \frac{1}{2}x_6^2$$

and

$$z_{2,2} = x_6 = y_2 - \dot{y}_1,$$

which shows that $z_{2,2}$, deduced via the applied diffeomorphism, contains the first derivative of y_1. This is indeed coherent with the fact that $dh_2(\bar{\tau}_{1,3}) \neq 0$ or $dL_f^2 h_2\left(\bar{\tau}_{1,1}\right) \neq 0$.

8.4 Diffeomorphism Deduction

Suppose that all conditions of Theorem 8.1 are satisfied, then we can easily extend the result on diffeomorphism deduction for dynamical systems with single output to treat the case with multiple outputs.

For this, let us evaluate the observability 1-forms defined in θ with the vector fields belonging to frame τ, which enables us to calculate the following matrix:

$$\begin{aligned} \Lambda &= \left(\theta_{j,k}(\tau_{i,l})_{1\le i\le m,\,1\le l\le r_i}\right)_{1\le j\le m,\,1\le k\le r_j} \\ &= \begin{pmatrix} \theta_{1,1}(\tau_{1,1}) & \cdots & \theta_{1,1}(\tau_{1,r_1}) & \theta_{1,1}(\tau_{2,1}) & \cdots & \theta_{1,1}(\tau_{m,r_m}) \\ \theta_{1,2}(\tau_{1,1}) & \cdots & \theta_{1,2}(\tau_{1,r_1}) & \theta_{1,2}(\tau_{2,1}) & \cdots & \theta_{1,2}(\tau_{m,r_m}) \\ \vdots & \ddots & \vdots & \vdots & \ddots & \vdots \\ \theta_{1,r_1}(\tau_{1,1}) & \cdots & \theta_{1,r_1}(\tau_{1,r_1}) & \theta_{1,r_1}(\tau_{2,1}) & \cdots & \theta_{1,r_1}(\tau_{m,r_m}) \\ \theta_{2,1}(\tau_{1,1}) & \cdots & \theta_{2,1}(\tau_{1,r_1}) & \theta_{2,1}(\tau_{2,1}) & \cdots & \theta_{2,1}(\tau_{m,r_m}) \\ \vdots & \ddots & \vdots & \vdots & \ddots & \vdots \\ \theta_{m,r_m}(\tau_{1,1}) & \cdots & \theta_{m,r_m}(\tau_{1,r_1}) & \theta_{m,r_m}(\tau_{2,1}) & \cdots & \theta_{m,r_m}(\tau_{m,r_m}) \end{pmatrix}. \end{aligned} \tag{8.16}$$

Since we have assumed that the observability rank condition imposed in Assumption 8.1 is satisfied, Λ defined in (8.16) is invertible. Denote Λ^{-1} as the inverse of Λ, then the following multi-valued 1-form can be computed

$$\omega = \Lambda^{-1}\theta, \tag{8.17}$$

where $\omega := (\omega_{i,j})_{1\leq i\leq m,\, 1\leq j\leq r_i}$.

Finally, if the components of the multi-valued 1-form ω are closed, i.e.,

$$d\omega_{i,j} = 0 \text{ for } 1 \leq i \leq m,\ 1 \leq j \leq r_i,$$

then, according to Poincaré's Lemma presented in Chap. 2, there exists a local diffeomorphism $\phi := (\phi_{i,j})_{1\leq i\leq m,\, 1\leq j\leq r_i}$ such that

$$\omega_{i,j} = d\phi_{i,j} \tag{8.18}$$

for $1 \leq i \leq m,\ 1 \leq j \leq r_i$.

Consequently, for the case with multiple outputs, the similar requirement that *the components of the multi-valued* $1\text{-}form$ ω *are closed* is a sufficient condition for the existence of the local diffeomorphism ϕ. As what we have seen in the previous chapters, if such a condition is satisfied, then we can construct a change of coordinates using (8.17).

8.5 Special Cases

In this section, we will state some special results that are similar to the single-output case. The first case is that the observability indices are the same and the second case is that their differences do not exceed 1 [2].

8.5.1 Equal Observability Indices

The first special case we analyze here is the case where the observability indices are equal.

Theorem 8.2 ([10]) *Assume that the observability indices of the dynamical system (8.1) are equal, i.e., the following condition holds:*

$$r_1 = r_2 = \cdots = r_m. \tag{8.19}$$

Then, the following two assertions are equivalent.

(1) The vector fields of the frame τ *commute, i.e.,*

$$\left[\tau_{i,k}, \tau_{j,s}\right] = 0$$

for all $1 \leq i,\ j \leq m$, $1 \leq k \leq r_i$ *and* $1 \leq s \leq r_j$;

(2) *There exists a local diffeomorphism* $z = \phi(x)$ *which transforms the dynamical system (8.1) into the normal form (8.2) where* φ *is the identity mapping on the output.*

The result stated in Theorem 8.2 is just a special case of Theorem 8.1, thus we can easily prove it by simply adapting the proof of Theorem 8.1. Consequently, according to Theorem 8.1, the item (1) can be considered as a criterion to verify the existence of a change of coordinates for system (8.1).

8.5.2 Unequal Observability Indices

In this section, we study the case where the observability indices are not equal, but satisfy the following inequality

$$|r_i - r_j| \leq 1$$

for any $1 \leq i, j \leq m$. In particular, we assume that

Assumption 8.2 For system (8.1), it is assumed that there exists an integer $\nu \in \{1, \ldots, m-1\}$ such that its observability indices satisfy the following properties:

- $r_1 = r_2 = \cdots = r_\nu = \rho_1$;
- $r_{\nu+1} = \cdots = r_m = \rho_2$;
- $\rho_1 = \rho_2 + 1$.

Under Assumption 8.2, in the following theorem, we study the existence of a change of coordinates for the dynamical system (8.1).

Theorem 8.3 *Assume that the observability indices of the dynamical system (8.1) satisfy Assumption 8.2. Then the following two assertions are equivalent*

(A_1) : *The vector fields of frame* τ *commute, i.e.,*

$$[\tau_{i,j}, \tau_{k,s}] = 0$$

for $1 \leq i \leq m$, $1 \leq j \leq r_i$, $1 \leq k \leq m$ *and* $1 \leq s \leq r_k$;

(A_2) : *The multi-valued 1-form* ω *given in (8.17) is closed.*

Moreover, if the assertion (A_1) *or* (A_2) *is fulfilled, then the following assertion is valid.*

(A_3) : *There exists a local diffeomorphism which transforms (8.1) into the normal form (8.2), whose the outputs are given by*

$$\begin{cases} \bar{y}_i = z_{i,r_j} = y_i, & \text{for } 1 \leq i \leq \nu, \\ \bar{y}_i = z_{i,r_i} = y_i + \gamma_j(y_1, \ldots, y_\nu), & \text{for } \nu+1 \leq i \leq m. \end{cases} \tag{8.20}$$

Proof Using the same argument in the previous section, it is easy to prove that (A_1) is equivalent to (A_2). In the following, we focus on proving that (A_3) can be deduced from (A_2).

For this, suppose that (A_2) is fulfilled, then from (8.18), we have $\phi_* := d\phi = \omega$, and we can deduce the following change of coordinates $z = \phi(x)$ with

$$z = (z_1^T, z_2^T, \ldots, z_m^T)^T$$

and

$$z_i = \left(z_{i,1}, z_{i,2}, \ldots, z_{i,r_i}\right)^T$$

for $1 \le i \le m$, where $dz_{i,j} = \omega_{i,j}$ for $1 \le i \le m$ and $1 \le j \le r_i$.

Now, let us calculate the transformation of the dynamical system (8.1) by using this change of coordinates. To this aim, compute the partial derivatives of $\phi_*(f)$ in the direction of $\dfrac{\partial}{\partial z_{i,j}}$ as follows:

$$\frac{\partial \phi_*(f)}{\partial z_{i,j}} = \left[\phi_*(f), \frac{\partial}{\partial z_{i,j}}\right] = \phi_*[f, \tau_{i,j}] = \omega(\tau_{i,j+1}) = \frac{\partial}{\partial z_{i+1}}$$

for $1 \le i \le m$ and $1 \le j \le r_i - 1$. Then, by integration, we obtain

$$\phi_*(f) = \sum_{j=1}^{m} \left(\sum_{k=1}^{r_j - 1} z_{j,k} \frac{\partial}{\partial z_{j,k+1}} \right) + \beta(z),$$

where β is the obtained function of z. Then it is clear that the linear part of $\phi_*(f)$ is of the form given in (8.2).

Finally, we seek the new outputs by calculating the Lie derivatives of the outputs h_k, for $1 \le k \le m$, along the vector fields $\left(\tau_{i,j}\right)_{1 \le i \le m,\, 1 \le j \le r_i}$ defined in (8.7)–(8.8). For $1 \le k \le \nu$, we obtain

$$\begin{cases} L_{\tau_{i,j}} h_k = 1, & \text{if } i = k,\ j = r_k \\ L_{\tau_{i,j}} h_k = 0, & \text{otherwise.} \end{cases}$$

Thus, we get the new outputs as

$$\bar{y}_k = y_k = z_{k,r_k}.$$

For $\nu + 1 \le k \le m$, we obtain

$$\begin{cases} L_{\tau_{i,j}} h_k = 1, & \text{if } i = k \text{ and } j = r_k \\ L_{\tau_{i,j}} h_k = 0, & \text{if } 1 \le i \le \nu \text{ and } j < r_i - 1 \\ L_{\tau_{i,j}} h_k = 0, & \text{if } \nu + 1 \le i \le m \text{ and } j \ne k. \end{cases}$$

If $1 \leq i \leq \nu$, then $L_{\tau_{i,r_i}} h_k$ might be different to 0. Hence, h_k can depend on y_i, for $1 \leq i \leq \nu$. Thus, we get the new outputs as

$$\bar{y}_k = z_{k,r_k} + \gamma(y_1, \ldots, y_\nu).$$

□

Different from the first special case where the observability indices are equal, we can have φ different from the identity mapping for the unequal case. This implies that we allow a diffeomorphism on the outputs. However, this does not influence the design of observers, as it contains only known variables.

The following example is to highlight the above discussion.

Example 8.2 Consider the following dynamical system

$$\begin{aligned}
\dot{x}_1 &= -k_1 x_1 + k_2 x_3 \sin(N_r x_4) + u_1 \\
\dot{x}_2 &= -k_1 x_2 - k_2 x_3 \cos(N_r x_4) + u_2 \\
\dot{x}_3 &= -k_4 x_3 - k_3 x_1 \sin(N_r x_4) - k_3 x_2 \cos(N_r x_4) \\
\dot{x}_4 &= x_3 \\
y_1 &= x_4, y_2 = x_2, y_3 = x_1
\end{aligned} \tag{8.21}$$

which describes the movement of a PM stepper motor [1].

For such a system, a straightforward calculation gives the following observability 1-forms:

$$\begin{aligned}
\theta_{1,1} &= dx_4, \theta_{1,2} = dx_3 \\
\theta_{2,1} &= dx_2, \theta_{3,1} = dx_1.
\end{aligned}$$

Secondly, based on (8.7)–(8.8), the vector fields forming the frame τ are given as follows:

$$\begin{aligned}
\tau_{1,1} &= \frac{\partial}{\partial x_3} \\
\tau_{1,2} &= \frac{\partial}{\partial x_4} + k_2 \sin(N_r x_4) \frac{\partial}{\partial x_1} - k_2 \cos(N_r x_4) \frac{\partial}{\partial x_2} - k_4 \frac{\partial}{\partial x_3} \\
\tau_{2,1} &= \frac{\partial}{\partial x_2}, \tau_{3,1} = \frac{\partial}{\partial x_1}.
\end{aligned}$$

Thirdly, an additional calculation gives

$$\Lambda = \theta\tau = \begin{pmatrix} 0 & 1 & 0 & 0 \\ 1 & -k_4 & 0 & 0 \\ 0 & -k_2 \cos(N_r x_4) & 1 & 0 \\ 0 & k_2 \sin(N_r x_4) & 0 & 1 \end{pmatrix}.$$

We can verify that $\tau_{1,1}$, $\tau_{1,2}$, $\tau_{2,1}$ and $\tau_{3,1}$ commute each other with respect to the Lie bracket, i.e., (A_1) holds. Moreover, the control directions $g_1(x) = \frac{\partial}{\partial x_1}$ and $g_2(x) =$

$\frac{\partial}{\partial x_2}$ do not depend on $\dot{y}_1$, since $[\tau_{1,1}, g_k] = 0$, for $k = 1, 2$. Please refer to Corollary 3.1 in Chap. 3 for more details.

Consequently, by calculating $\omega = \Lambda^{-1}\theta$ defined in (8.17), we get

$$\omega = \begin{pmatrix} k_4 dx_4 + dx_3 \\ dx_4 \\ k_2 \cos(x_4 N_r) dx_4 + dx_2 \\ -k_2 \sin(x_4 N_r) dx_4 + dx_1 \end{pmatrix},$$

which gives the following diffeomorphism:

$$\begin{pmatrix} z_{1,1} \\ z_{1,2} \\ z_{2,1} \\ z_{3,1} \end{pmatrix} = \begin{pmatrix} k_4 x_4 + x_3 \\ x_4 \\ x_2 + k_2 \sin x_4 N_r \\ x_1 + k_2 \cos x_4 N_r \end{pmatrix}.$$

Finally, using Theorem 8.3 we obtain the following form:

$$\begin{aligned} \dot{z}_{1,1} &= -k_3 y_1 \sin(N_r y_3) - k_3 y_2 \cos(N_r y_3) \\ \dot{z}_{1,2} &= z_{1,1} - k_4 y_3 \\ \dot{z}_{2,1} &= -k_1 y_2 + u_2 \\ \dot{z}_{3,1} &= -k_1 y_1 + u_1 \\ \bar{y}_1 &= y_1 + \tfrac{k_2}{N_r} \cos(N_r y_3) = z_{3,1} \\ \bar{y}_2 &= y_2 + \tfrac{k_2}{N_r} \sin(N_r y_3) = z_{2,1} \\ \bar{y}_3 &= y_3 = z_{1,2}. \end{aligned}$$

8.6 Extension to Partial Observer Normal Form with Multiple Outputs

This section[1] is dedicated to analyzing the partial observer normal form for system (8.1). For this, the following assumption is imposed.

Assumption 8.3 For system (8.1), it is assumed that the observability indices $\{r_i\}_{1 \le i \le m}$ are constant on the open set $\mathscr{X}$, and satisfy

$$\sum_{i=1}^{m} r_i = r < n.$$

[1] Section 8.6 contains excerpts from [8]. from Saadi, W., Boutat, D., Zheng, G., Sbita, L., Yu, L.: Algorithm to compute nonlinear partial observer normal form with multiple outputs. IEEE Transactions on Automatic Control 65(6), 2700–2707 (2020).

Under Assumption 8.3 where the studied system (8.1) is only partially observable, the targeted observer normal form has to be adapted into the following one

$$\dot{z}_i = A_{O_i} z_i + \beta_i(\bar{y}) \tag{8.22}$$
$$\dot{\eta} = \xi(z, \eta) \tag{8.23}$$
$$\bar{y}_i = C_{O_i} z_i = z_{i,r_i} = y_i + \varphi(y_1, \ldots, y_{i-1}), \tag{8.24}$$

where A_{O_i} and C_{O_i} are respectively the $(r_i \times r_i)$ and $(1 \times r_i)$ well-known Brunovsky forms defined in (8.3). The objective is then to deduce necessary and sufficient geometric conditions that guarantee the existence of a diffeomorphism $\phi(x) = (z^T, \eta^T)^T$ where $z = \left(z_1^T, \ldots, z_m^T\right)^T$ with $z_i = \left(z_{i,1}, \ldots, z_{i,r_i}\right)^T$ for $1 \leq i \leq m$ and $\eta = (\eta_1, \ldots, \eta_{n-r})^T$ such that (8.1) might be transformed into (8.24).

For (8.1), as we have presented in Sect. 8.2, we can firstly calculate $\theta_{i,j} = dL_f^{j-1} h_i$, and then construct the corresponding co-distribution Δ as follows:

$$\Delta = \text{span}\left\{\theta_{i,j}, \text{ for } 1 \leq i \leq m \text{ and } 1 \leq j \leq r_i\right\}. \tag{8.25}$$

Under Assumption 8.3, the above co-distribution is of dimension r where

$$r = \text{rank} = \sum_{\mathrm{i=1}}^{\mathrm{m}} \mathrm{r_i} \leq \mathrm{n},$$

which implies that $dL_f^k h_j \in \Delta$ for any positive integer k.

In this case, we can follow the same method when treating partial observer normal form for systems with single output in Chap. 7 to determine the annihilator (or kernel) of Δ as follows:

$$\Delta^{\perp} = \{X : \nu(X) = 0, \quad \text{for any} \quad \nu \in \Delta\}, \tag{8.26}$$

which is a distribution of dimension $n - r$.

In order to construct the change of coordinates for the partially observable case, like what we did for the fully observable case, we need as well to introduce a family of vector fields τ. Here, we follow the same way for the full observable case to build such a frame, i.e., the first m vector fields $\bar{\tau}_{i,1}$ for $1 \leq i \leq m$ are chosen as one solution of the algebraic equations (8.7), and calculate the rest of vector fields by using (8.8).

8.6.1 Properties of Δ and $\Delta^{\perp}$

When treating partial observer normal form for systems with single output, we have stated in Chap. 7 that the co-distribution Δ and the distribution $\Delta^{\perp}$ process certain properties (see Lemma 7.1). The following will investigate similar properties for multiple-output case.

Lemma 8.1 *Consider the co-distribution Δ defined in (8.25) and the distribution $\Delta^\perp$ defined in (8.26) for system (8.1), we have the following properties:*

(1) Δ and $\Delta^\perp$ are involutive;
(2) Δ and $\Delta^\perp$ are invariant with respect to f of (8.1) in the sense that

$$L_f \nu \in \Delta, \; \forall \nu \in \Delta$$

and

$$[f, H] \in \Delta^\perp, \; \forall H \in \Delta^\perp;$$

(3) there exist $(n - r)$ vector fields, noted as $\{\tau_{r+1}, \ldots, \tau_n\}$, that span $\Delta^\perp$ and commute, i.e.,

$$[\tau_i, \tau_j] = 0 \textit{ for } r + 1 \le i, j \le n.$$

Proof Let us firstly prove the first item of Lemma 8.1. Since the co-distribution Δ defined in (8.25) is spanned by the exact forms $\theta_{j,k}$, therefore it is involutive. Besides, for any two vector fields $X, Y \in \Delta^\perp$, according to (8.15), we can then get

$$d\theta_{i,j} = 0$$

and

$$\theta_{i,j}(Y) = \theta_{i,j}(X) = 0$$

for $1 \le i \le m$ and $1 \le j \le r_i$. This yields $\theta_{i,j}([X, Y]) = 0$, which is equivalent to

$$[X, Y] \in \Delta^\perp.$$

Hence, $\Delta^\perp$ is also involutive, and we proved (1) of Lemma 8.1.

Concerning the second item of Lemma 8.1, by applying the well-known Cartan's identity presented in Chap. 2, we get

$$L_f\theta_{i,j} = \imath_f d\theta_{i,j} + d\imath_f\theta_{i,j} = d\imath_f\theta_{i,j}$$

because $d\theta_{i,j} = 0$. Moreover, since Assumption 8.3 is supposed to be fulfilled, i.e., the partial observability rank condition is satisfied, then we obtain

$$d\imath_f\theta_{i,j} = d\theta_{i,j}(f) = \theta_{i+1}$$

for $1 \le j \le r_i - 1$ and

$$d\theta_{i,r_i}(f) = dL_f^{r_i+1}\theta_{i,r_i} \in \Delta.$$

Thus, we always have the following equation:

$$L_f\theta_{i,j} \in \Delta.$$

Now, let us choose $H \in \Delta^{\perp}$. Due to the fact that $\theta_{i,j}$ for $1 \leq i \leq m$ and $1 \leq j \leq r_i$ are closed, by applying (8.15), we can get

$$\theta_{i,j}([f, H]) = L_f \theta_{i,j}(H) - L_H \theta_{i,j}(f) = -L_H \theta_{i,j}(f).$$

Note that the interior product of a function by a vector field is equal to zero, thus $i_H \theta_{i,j}(f) = 0$. Based on this result, and by applying again the Cartan's identity, we get

$$L_H \theta_{i,j}(f) = i_H d\theta_{i,j}(f).$$

As $d\theta_{i,j}(f) = L_f \theta_{i,j} \in \Delta^{\perp}$, we obtain

$$\theta_{i,j}([f, H]) = 0,$$

thus $[f, H] \in \Delta^{\perp}$, and we proved (2) of Lemma 8.1.

Finally, since $\Delta^{\perp}$ is involutive, it is well-known that we can find $(n - r)$ vector fields which form a basis $\{\tau_{r+1}, \ldots, \tau_n\}$, such that $[\tau_i, \tau_j] = 0$ for all $r + 1 \leq i, j \leq n$. Therefore, we proved (3) of Lemma 8.1.

□

Lemma 8.2 *For system (8.1) with a family of vector fields $\tau_{i,j}$ deduced by (8.7)–(8.8) for $1 \leq i \leq m$ and $1 \leq j \leq r_i$, $\Delta^{\perp}$ defined in (8.26) is invariant with respect to $\tau_{i,j}$ in the sense that*

$$[\tau_{i,j}, H] \in \Delta^{\perp} \tag{8.27}$$

for any $H \in \Delta^{\perp}$.

Proof In fact, we need only to prove (8.27) is true for $\tau_{i,1}$ with $1 \leq i \leq m$. To show this, let us calculate $\tau_{i,j} = [\tau_{i,j-1}, f]$ for $1 \leq i \leq m$ and $1 \leq j \leq r_i$, and then use the well-known Jacobi's identity (2.15), which enables us to obtain

$$[\tau_{i,j}, H] = [[\tau_{i,j-1}, f], H] = -[[H, \tau_{i,j-1}], f] - [[f, H], \tau_{i,j-1}].$$

Based on (2) of Lemma 8.1 (i.e., $[f, H] \in \Delta^{\perp}$), it is clear to see that $[\tau_{i,j}, H]$ belongs to $\Delta^{\perp}$ if and only if $[\tau_{i,j-1}, H] \in \Delta^{\perp}$. Hence, if we can prove (8.27) is true for $\tau_{i,1}$ with $1 \leq i \leq m$, then it will be satisfied as well for $\tau_{i,j}$ with $1 \leq i \leq m$ and $2 \leq j \leq r_i$. Thus, the following is to prove (8.27) is true for $\tau_{i,1}$ with $1 \leq i \leq m$.

For this, we need to prove

$$\theta_{s,t}(\left[\tau_{i,1}, H\right]) = 0$$

for $1 \leq i \leq m$, $1 \leq s \leq m$ and $1 \leq t \leq r_s$, and this is equivalent to $\left[\tau_{i,1}, H\right] \in \Delta^{\perp}$.

As $\theta_{s,t}$ for $1 \leq s \leq m$ and $1 \leq t \leq r_s$ are closed and $\theta_{s,t}(H) = 0$ for any $H \in \Delta^{\perp}$, then from (8.15) we have

$$\theta_{s,t}(\left[\tau_{i,1}, H\right]) = L_H \theta_{s,t}(\tau_{i,1})$$

for which the following two different cases should be distinguished.
Case 1: $i = 1$
According to (8.7), we know that $\theta_{s,t}(\tau_{1,1})$ is a constant (either 0 or 1 for different indices), therefore, its Lie derivative is 0. Thus, we obtain

$$\theta_{s,t}([\tau_{1,1}, H]) = L_H\theta_{s,t}(\tau_{1,1}) = 0,$$

which implies that $[\tau_{1,1}, H] \in \Delta^{\perp}$.
Case 2: $2 \leq i \leq m$
For $i \neq 1$, the function $\theta_{s,t}(\tau_{i,1})$ is not necessary to be constant for all $\theta_{s,t}$. In what follows, we will prove the result for $i = 2$, and the same argument can be used to prove the result for $3 \leq i \leq m$.

According to (8.7), it can be seen that $\theta_{s,t}(\tau_{2,1}) = 0$ for $2 \leq s \leq m$ and $1 \leq t \leq r_s$, except when $s = 2$ and $t = r_2$ for which $\theta_{2,r_2}(\tau_{2,1}) = 1$. On the other hand, we have $\theta_{1,t}(\tau_{2,1}) = 0$ for $1 \leq t \leq r_2$. Consequently, for all $\theta_{s,t}$, we have

$$\theta_{s,t}([\tau_{2,1}, H]) = 0$$

for any $H \in \Delta^{\perp}$.

It remains us to prove that the result is true as well for $\theta_{1,r_2+1}, \ldots, \theta_{1,r_1}$. As we have proved in the above that it is always true if $\theta_{1,k}(\tau_{2,1})$ for $r_2 + 1 \leq k \leq r_1$ is constant. In the following, we will show that it is true as well even if $\theta_{1,k}(\tau_{2,1})$ is not constant.

For this, for any $\theta_{1,k}(\tau_{2,1}) = \lambda$ which is not constant, we can rewrite it as follows:

$$\theta_{1,k} = \lambda\theta_{2,r_2} + \nu, \tag{8.28}$$

where ν is 1-form such that $\nu(\tau_{2,1}) = 0$ since $\theta_{1,k}$ and θ_{2,r_2} belong to Δ which implies $\nu \in \Delta$. Therefore, we have

$$\nu = \sum q_{i,j}\theta_{i,j}, \tag{8.29}$$

which is a combination of some $\theta_{i,j}$ that vanish on $\bar{\tau}_{2,1}$ and $q_{i,j}$ are functions.

As $\theta_{2,r_2}([\tau_{2,1}, H]) = 0$ because of (8.15), due to the fact that $\theta_{2,r_2}(\tau_{2,1}) = \theta_{2,r_2}(H) = 0$, then $\theta_{2,r_2}([\tau_{2,1}, H]) = 0$ is equivalent to $\nu([\tau_{2,1}, H]) = 0$. To show this, consider again (8.15) for ν as follows:

$$d\nu(\tau_{2,1}, H) = L_{\tau_{2,1}}\nu(H) - L_H\nu(\tau_{2,1}) - \nu([\tau_{2,1}, H]). \tag{8.30}$$

As $\nu(H) = \nu(\tau_{2,1}) = 0$, this equation becomes

$$d\nu(\tau_{2,1}, H) = -\nu([\tau_{2,1}, H]).$$

On the other hand, by differentiation of (8.29), we obtain

$$d\nu = \sum dq_{i,j} \wedge \theta_{i,j} + \sum q_{i,j} d\theta_{i,j} = \sum dq_{i,j} \wedge \theta_{i,j}$$

since $d\theta_{i,j} = 0$. Therefore,

$$d\nu(\tau_{2,1}, H) = \sum dq_{i,j}(\tau_{2,1})\theta_{i,j}(H) - dq_{i,j}(H)\theta_{i,j}(\tau_{2,1}) = 0$$

since $\theta_{i,j}(H) = \theta_{i,j}(\tau_{2,1}) = 0$. Finally, we obtain

$$\theta_{1,k}([\tau_{2,1}, H]) = \nu([\tau_{2,1}, H]) = 0.$$

Thus, we proved that $[\tau_{2,1}, H] \in \Delta^{\perp}$ for any $H \in \Delta^{\perp}$. □

Based on the above lemma, we can then state the following result.

Corollary 8.1 *For system (8.1) with a family of vector fields $\tau_{i,j}$ deduced by (8.7)–(8.8) for $1 \leq i \leq m$ and $1 \leq j \leq r_i$, for any $H_i \in \Delta^{\perp}$ and $H_j \in \Delta^{\perp}$ we have*

$$\left[\tau_{i,k} + H_i,\ \tau_{j,s} + H_j\right] = \left[\tau_{i,k},\ \tau_{j,s}\right] \text{ modulo } \Delta^{\perp}. \tag{8.31}$$

Proof Given a family of vector fields $\tau_{i,j}$ deduced by Eqs. (8.7)–(8.8) for $1 \leq i \leq m$ and $1 \leq j \leq r_i$, let us calculate the following Lie bracket

$$\left[\tau_{i,k} + H_i, \tau_{j,s} + H_j\right] = \left[\tau_{i,k}, \tau_{j,s}\right] + \left[\tau_{i,k}, H_j\right] + \left[H_i, \tau_{j,s}\right] + \left[H_i, H_j\right].$$

As we have proved in Lemma 8.1 that $\Delta^{\perp}$ is involutive, therefore

$$\left[H_i, H_j\right] \in \Delta^{\perp}.$$

From Lemma 8.2, we deduced that $\left[\tau_{i,k}, H_j\right]$ and $\left[H_i, \tau_{j,s}\right]$ both belong to $\Delta^{\perp}$, thus we proved the result stated in Corollary 8.1. □

8.6.2 *Transformation*

This section deals with the geometric condition that guarantees the existence of a change of coordinates transforming system (8.1) into the proposed observer normal form (8.22)–(8.24).

Theorem 8.4 *There exists a local diffeomorphism $(z^T, \eta^T)^T = \phi(x)$ via which (8.1) can be transformed into (8.22)–(8.24) if and only if the following conditions are satisfied*

(1) there exists a family of vector fields $\tau_{i,1}$ for $1 \leq i \leq m$ which are solutions of algebraic equations (8.7), we have

$$\left[\tau_{i,k}, \tau_{j,s}\right] \in \Delta^{\perp}, \tag{8.32}$$

where $\tau_{i,k}$ *and* $\tau_{j,s}$ *are defined by (8.8) for all* $1 \leq i, j \leq m$, $1 \leq k \leq r_i$ *and* $1 \leq s \leq r_j$;

(2) *for all* $1 \leq i \leq m$, $1 \leq k \leq r_i$ *we have*

$$\left[\tau_{i,k}, \bar{\tau}_l\right] \in \Delta, \tag{8.33}$$

where the vector fields $\bar{\tau}_l$, *for* $r+1 \leq l \leq n$, *are deduced in the item* (3) *of Lemma 8.1;*

(3) *for* $1 < i < j \leq m$ *and* $r_i > r_j$ *we have*

$$dh_j(\tau_{i,r_j+k}) = 0 \text{ and } dh_j(\tau_{s,r_s-(r_i+t)+r_j+k}) = 0, \tag{8.34}$$

with $1 \leq k \leq r_i - r_j$, $1 \leq s \leq i-1$ *and* $0 \leq t \leq r_s - r_i$.

Before to prove Theorem 8.4, the following statements can be made by applying the results of Lemma 8.2 and Corollary 8.1.

Remark 8.1 We have the following equivalences:

- The items (1), (2) and (3) of Theorem 8.4 are fulfilled by at least one family of vector fields $\tau_{i,j}$.
- The items (1), (2) and (3) of Theorem 8.4 are fulfilled for the family of vector fields $\bar{\tau}_{i,j}$ obtained from any $\tau_{i,j}$ by removing the components in $\bar{\tau}_l$ directions. Especially, this family fulfills the following commutativity conditions $\left[\bar{\tau}_{i,k}, \bar{\tau}_{j,s}\right] = 0$ and $\left[\bar{\tau}_{i,k}, \bar{\tau}_l\right] = 0$.

Let us show an example to highlight the first two items in Remark 8.1.

Example 8.3 Consider the following dynamical system:

$$\begin{aligned}
\dot{x}_1 &= x_3 x_1 \\
\dot{x}_2 &= x_1 \\
\dot{x}_3 &= -x_3 + x_3 x_2^2 \\
\dot{x}_4 &= -x_4 + x_1 x_3^2 + x_4^2 \\
y_1 &= x_2,\ y_2 = x_3.
\end{aligned}$$

Let us firstly calculate the corresponding 1-forms

$$\begin{aligned}
\theta_{1,1} &= dx_2 \\
\theta_{1,2} &= dx_1 \\
\theta_{2,1} &= dx_3.
\end{aligned}$$

It is easy to see that the above obtained 1-forms are linearly independent. Hence, according to (8.5), we get

$$\Delta = \text{span}\left\{\theta_{1,1}, \theta_{1,2}, \theta_{2,1}\right\},$$

which is of rank 3. Based on Definition 8.1, we obtain the related observability indices as $r_1 = 2$ and $r_2 = 1$. Moreover, according to (8.26), we get

$$\Delta^{\perp} = \text{span}\left\{\frac{\partial}{\partial x_4}\right\}.$$

Moreover, by solving (8.7), we can choose the following solution:

$$\begin{aligned} \tau_{11} &= \frac{\partial}{\partial x_1} \\ \tau_{12} &= \frac{\partial}{\partial x_2} + x_3\frac{\partial}{\partial x_1} + x_3^2\frac{\partial}{\partial x_4} \\ \tau_{13} &= \frac{\partial}{\partial x_3} \end{aligned}$$

and it is clear that

$$[\tau_{12}, \tau_{13}] = -\frac{\partial}{\partial x_1} - 2x_3\frac{\partial}{\partial x_4} \notin \Delta^{\perp}.$$

However, if we choose

$$\tau_{13} = \frac{\partial}{\partial x_3} + x_2\frac{\partial}{\partial x_1},$$

then we obtain the following Lie bracket:

$$[\tau_{12}, \tau_{13}] = -2x_3\frac{\partial}{\partial x_4} \in \Delta^{\perp}.$$

As $\bar{\tau}_4 = \frac{\partial}{\partial x_4}$, then (8.33) in Theorem 8.4 is satisfied. In fact, the second item of Remark 8.1 implies that we should take

$$\tau_{12} = \frac{\partial}{\partial x_2} + x_3\frac{\partial}{\partial x_1} \text{ modulo } \Delta^{\perp}.$$

The last condition in Theorem 8.4 is obviously satisfied since we have $r_1 = 2$ and $r_2 = 1$.

In summary, the above discussions are to highlight the fact that if we take the following diffeomorphism

$$\begin{aligned} z_1 &= x_1 - x_3x_2 \\ z_2 &= x_2, z_3 = x_3, z_4 = x_4, \end{aligned}$$

then we obtain the following partial observable normal form:

$$\begin{aligned} \dot{z}_1 &= z_2z_3(1 - z_2^2) \\ \dot{z}_2 &= z_1 + z_2z_3 \\ \dot{z}_3 &= -z_3 + z_3z_2^2 \\ \dot{z}_4 &= -z_4 + z_1z_3^2 + z_3z_4^2 + z_4^2 \\ y_1 &= z_2, y_2 = z_3. \end{aligned}$$

The proof of Theorem 8.4 is based on the following lemma.

Lemma 8.3 *For a family of vector fields $\tau_{1,j}$ satisfying (8.7) and the associated $\tau_{i,j}$ for $2 \le j \le r_i$ deduced from (8.8), there exists a family of vector fields*

$$\bar{\tau}_{i,j} = \tau_{i,j} \text{ modulo } \Delta^{\perp}$$

such that $\left[\bar{\tau}_{i,j}, \bar{\tau}_{s,t}\right] = 0$ for $1 \le j \le r_i$ and $1 \le t \le r_s$, if and only if

$$[\tau_{i,j}, \tau_{s,t}] \in \Delta^{\perp}.$$

Proof **Necessity:** Suppose that there exist $H_{i,j} \in \Delta^{\perp}$ and $H_{s,t} \in \Delta^{\perp}$, which determine $\bar{\tau}_{i,j} = \tau_{i,j} + H_{i,j}$ and $\bar{\tau}_{s,t} = \tau_{s,t} + H_{s,t}$ for $1 \le j \le r_i$ and $1 \le t \le r_s$ such that $\left[\bar{\tau}_{i,j}, \bar{\tau}_{s,t}\right] = 0$, then we have

$$0 = \left[\bar{\tau}_{i,j}, \bar{\tau}_{s,t}\right] = [\tau_{i,j}, \tau_{s,t}] \text{ modulo } \Delta^{\perp},$$

which implies that $[\tau_{i,j}, \tau_{s,t}] \in \Delta^{\perp}$.

Sufficiency: Assume that for one choice of the family of vector fields $\tau_{1,j}$ satisfying (8.7), we have $[\tau_{i,j}, \tau_{s,t}] \in \Delta^{\perp}$ for $1 \le j \le r_i$ and $1 \le t \le r_s$ defined in (8.8). Then we need to prove that there exist $H_{i,j} \in \Delta^{\perp}$, $H_{s,t} \in \Delta^{\perp}$ for $1 \le j \le r_i$ and $1 \le t \le r_s$ such that $[\bar{\tau}_{i,j}, \bar{\tau}_{s,t}] = 0$ where $\bar{\tau}_{i,j} = \tau_{i,j} + H_{i,j}$ and $\bar{\tau}_{s,t} = \tau_{s,t} + H_{s,t}$.

Let $\{\bar{\tau}_{r+1}, \ldots, \bar{\tau}_n\}$ be $(n-r)$ determined commutative basis of $\Delta^{\perp}$, it is well known that there exist r commutative vector fields, noted as $\{X_1, \ldots, X_r\}$, such that they commute with $\{\bar{\tau}_{r+1}, \ldots, \bar{\tau}_n\}$, thus form together a basis of the tangent fiber bundle $T\mathscr{X}$ of $\mathscr{X}$.

Therefore, we can rewrite $\tau_{i,j}$ as

$$\tau_{i,j} = \sum_{l=1}^{r} \gamma_l^{i,j}(x) X_l + \sum_{l=r+1}^{n} \mu_l^{i,j}(x)\bar{\tau}_l,$$

where functions $\gamma_l^{i,j}(x)$ are constant on the leaves of $\Delta^{\perp}$, thus $L_{\bar{\tau}_i}\gamma_l^{i,j} = 0$ because of the invariance of $\Delta^{\perp}$ by means of $\tau_{i,j}$ thanks to Lemma 8.2, i.e., $[\tau_{i,j}, \bar{\tau}_i] \in \Delta^{\perp}$. By choosing

$$H_{i,j} = -\sum_{l=r+1}^{n} \mu_l^{i,j}(x)\bar{\tau}_l \in \Delta^{\perp},$$

we have

$$\bar{\tau}_{i,j} = \tau_{i,j} + H_{i,j} = \sum_{l=1}^{r} \gamma_l^{i,j}(x) X_l.$$

For the same reason, there exists $H_{s,t} = \sum_{l=r+1}^{n} \kappa_l^{s,t}(x)\bar{\tau}_l \in \Delta^{\perp}$ yielding

$$\bar{\tau}_{s,t} = \tau_{s,t} + H_{s,t} = \sum_{l=1}^{r} \alpha_{j,l}(x) X_l.$$

Moreover, for $1 \leq j \leq r_i$ and $1 \leq s \leq r_t$, due to the invariance of $\Delta^{\perp}$ by means of $\tau_{i,j}$, we have

$$[\tau_{i,j}, \tau_{s,t}] = [\bar{\tau}_{i,j}, \bar{\tau}_{s,t}] \text{ modulo } \Delta^{\perp}. \tag{8.35}$$

As $[X_k, X_s] = 0$ for $1 \leq k, s \leq r$ and the fact that functions $\gamma_l^{i,j}$ and $\alpha_l^{i,j}$ are constant on the leaves of $\Delta^{\perp}$, then (8.35) becomes

$$[\tau_i, \tau_j] = \left[\sum_{l,m=1}^{r} \gamma_l^{i,j} L_{X_l} \alpha_m^{s,t} X_s - \sum_{m,l=1}^{r} \alpha_l^{s,t} L_{X_m} \gamma_l^{i,j} X_l \right] \text{ modulo } \Delta^{\perp}.$$

Thus, if $[\tau_{i,j}, \tau_{s,t}] \in \Delta^{\perp}$, then there exist $H_{i,j} \in \Delta^{\perp}$ and $H_{s,t} \in \Delta^{\perp}$ defined above, such that

$$[\bar{\tau}_{i,j}, \bar{\tau}_{s,t}] = \sum_{l,m=1}^{r} \gamma_l^{i,j} L_{X_l} \alpha_m^{s,t} X_s - \sum_{m,l=1}^{r} \alpha_l^{s,t} L_{X_m} \gamma_l^{i,j} X_l = 0,$$

which is due to the fact that $X_k \notin \Delta^{\perp}$. □

Using the result stated in Lemma 8.3, the proof of Theorem 8.4 is detailed in the following.

Proof (*of Theorem* 8.4.) **Necessity:** Suppose that there exists a local change of coordinates $(z^T, \eta^T)^T = \phi(x)$ which transforms (8.1) into (8.22)–(8.24). Note $\omega = d\phi(x)$, then ω is the pullback of dz and $d\eta$ for (8.22)–(8.23). Therefore, we need only to check the necessity for (8.22)–(8.23). It is clear that, for (8.22)–(8.23), there exist

$$\bar{\tau}_{i,k} = \frac{\partial}{\partial z_{i,k}}$$

for $1 \leq i \leq m$ and $1 \leq k \leq r_i$ and

$$\bar{\tau}_j = \frac{\partial}{\partial \eta_j}$$

for $1 \leq j \leq n - r$ such that all conditions in Theorem 8.4 are satisfied for (8.22)–(8.23). Then, according to the property of pullback, we can conclude that those conditions are necessary for (8.1).

Sufficiency: Suppose that all conditions in Theorem 8.4 are satisfied. Then we can construct the following multi-valued matrix Λ

$$
\Lambda = \begin{pmatrix}
\theta_{1,1}(\tau_{1,1}) & \cdots & \theta_{1,1}(\tau_{1,r_1}) & \theta_{1,1}(\tau_{2,1}) & \cdots & \theta_{1,1}(\tau_{m,r_m}) \\
\vdots & \ddots & \vdots & \vdots & \ddots & \vdots \\
\theta_{1,r_1}(\tau_{1,1}) & \cdots & \theta_{1,r_1}(\tau_{1,r_1}) & \theta_{1,r_1}(\tau_{2,1}) & \cdots & \theta_{1,r_1}(\tau_{m,r_m}) \\
\theta_{2,1}(\tau_{1,1}) & \cdots & \theta_{2,1}(\tau_{1,r_1}) & \theta_{2,1}(\tau_{2,1}) & \cdots & \theta_{2,1}(\tau_{m,r_m}) \\
\vdots & \ddots & \vdots & \vdots & \ddots & \vdots \\
\theta_{m,r_m}(\tau_{1,1}) & \cdots & \theta_{m,r_m}(\tau_{1,r_1}) & \theta_{m,r_m}(\tau_{2,1}) & \cdots & \theta_{m,r_m}(\tau_{m,r_m})
\end{pmatrix}.
$$

As the 1-forms θ_{ij} are independent, so are the vector fields. As $\tau_{st} \notin \Delta^{\perp}$ where $\Delta^{\perp}$ is the kernel of θ_{ij}, then Λ is invertible. Thus, we can define r differential 1-forms as follows:

$$
\omega = \begin{pmatrix} \omega_{1,1} \\ \vdots \\ \omega_{1,r_1} \\ \omega_{2,1} \\ \vdots \\ \omega_{m,r_m} \end{pmatrix} = \Lambda^{-1} \begin{pmatrix} \theta_{1,1} \\ \vdots \\ \theta_{1,r_1} \\ \theta_{2,1} \\ \vdots \\ \theta_{m,r_m} \end{pmatrix}. \tag{8.36}
$$

From the above formula, we see that 1-forms $\omega_{i,j}$ are linear combinations of 1-forms $\theta_{i,j}$. Thus, they span the same co-distribution Δ and have the annihilator as $\Delta^{\perp}$. Therefore, they are well defined from any family $\tau_{i,1}$ modulo $\Delta^{\perp}$. By the definition we have

$$
\omega_{i,j}(\tau_{s,t}) = \delta_i^s \delta_j^t,
$$

where δ_i^s is the Kronecker delta, therefore if the condition (8.32) of Theorem 8.4 is fulfilled, then we can find $n-r$ differential 1-forms ω_j for $1 \leq j \leq n-r$ such that

- $\omega_j(\bar{\tau}_k) = \delta_k^j$ for $r+1 \leq k, j \leq n$;
- $\omega_j(\tau_{s,t}) = 0$ thus the annihilator of these 1-forms is spanned by $\tau_{s,t}$.

Note that, by using (8.15), the above 1-forms ω_j together with $\omega_{i,k}$ are independent and closed. Then, according to the construction procedure, we have

$$
\omega_j(\bar{\tau}_j) = \frac{\partial}{\partial \eta_j}
$$

and

$$
\omega_{i,k}(\tau_{i,k}) = \frac{\partial}{\partial z_{i,k}}.
$$

Therefore, thanks to Poincaré's Lemma, there exists a change of coordinates $\phi(x) = (z^T, \eta^T)^T$ such that the differential of ϕ is $\phi_* = \omega$.

Let us show how the vector field f in the dynamical system (8.1) is transformed by ϕ. To do so, we compute $\frac{\partial}{\partial \eta_j}\phi_*(f)$, and we obtain

$$\frac{\partial}{\partial \eta_j}\phi_*(f) = \left[\frac{\partial}{\partial \eta_j}, \phi_*(f)\right].$$

Then we have

$$\frac{\partial}{\partial \eta_j}\phi_*(f) = \phi_*([\bar{\tau}_j, f]) \in \phi_*(\Delta^\perp)$$

because $\bar{\tau}_j \in \Delta^\perp$ and $\Delta^\perp$ is f-invariant according to Lemma 8.1. Therefore we get

$$\frac{\partial}{\partial \eta_j}\phi_*(f) \in \operatorname{span}\left\{\frac{\partial}{\partial \eta_k}, \text{ for } r+1 \leq k \leq n\right\}.$$

Now we compute $\frac{\partial}{\partial z_{i,j}}\phi_*(f)$ for $1 \leq i \leq m$ and $1 \leq j \leq r_i - 1$, and we obtain

$$\begin{aligned}\tfrac{\partial}{\partial z_{i,j}}\phi_*(f) &= \phi_*([\tau_{i,j}, f]) = \phi_*(\tau_{i,j+1} + H)\\ &= \phi_*(\tau_{i,j+1}) + \phi_*(H),\end{aligned}$$

where $H \in \Delta^\perp$ which gives

$$\frac{\partial}{\partial z_{i,j}}\phi_*(f) = \frac{\partial}{\partial z_{i,j+1}} \text{ modulo } \Delta^\perp.$$

Therefore, by integration, we have

$$\phi_*(f) = \sum_{j=1}^{m}\left(\sum_{k=1}^{r_j-1} z_{j,k}\frac{\partial}{\partial z_{j,k+1}}\right) + \beta\left(z_{r_1}, \ldots, z_{r_m}\right) \text{ modulo } \Delta^\perp.$$

It rests to find the relation between $\left(z_{r_1}, \ldots, z_{r_m}\right)$ and the outputs $y_1, \ldots, y_m$. For this, it should be noted that from (8.34), together with (8.7) for $1 \leq i, j \leq m$ and $r_i > r_j + 1$, we have $dL_f^k h_j(\tau_{i,1}) = 0$ for $0 \leq k \leq r_i - 1$.

Now, using (8.15) for $\nu = dL_f^k h_j$, $X = f$ and $Y = \tau_{i,1}$, we obtain

$$0 = L_f dL_f^k h_j(\tau_{i,1}) - L_{\bar{\tau}_{i,1}} dL_f^k h_j(f) - dL_f^k h_j[f, \tau_{i,1}].$$

If $dL_f^k h_j(\tau_{i,1}) = 0$, then we obtain

$$dL_f^{k+1} h_j(\tau_{i,1}) = dL_f^k h_j(\tau_{i,2}).$$

Therefore, by induction, we deduce that

$$dh_j(\tau_{i,r_i-1}) = \cdots = dh_j(\tau_{i,1}) = 0.$$

Thus, the output h_j does not depend on $\dot{y}_i, \ldots, y_i^{(r_i-1)}$. However, $dh_j(\tau_{i,r_i})$ may not be equal to 0. Therefore, by integration, we obtain

$$y_j = z_{r_j} + \varphi_j(z_{r_1}, \ldots, z_{r_{j-1}}).$$

As $z_{r_1} = y_1$, $z_{r_2} = y_2 + \varphi(y_1)$ and so forth $z_{r_m} = y_m + \varphi(y_1, \ldots, y_{m-1})$, finally, we obtain (8.24). □

In order to highlight the proposed result, let us consider an example.

Example 8.4 Consider the following nonlinear system:

$$\begin{aligned}
\dot{x}_1 &= x_2x_5 + x_3x_4 - x_3^3 + x_3x_5 \\
\dot{x}_2 &= x_1, \dot{x}_3 = x_2 \\
\dot{x}_4 &= -2x_3x_4 + 2x_3^3 - 2x_2x_5 \\
\dot{x}_5 &= x_4 - x_3^2 \\
\dot{x}_6 &= -x_6 + x_2x_5 \\
y_1 &= x_3, y_2 = x_5.
\end{aligned} \tag{8.37}$$

According to the presented scheme, let us firstly calculate the corresponding 1-forms

$$\begin{aligned}
\theta_{1,1} &= dx_3, \theta_{1,2} = dx_2 \\
\theta_{1,3} &= dx_1, \theta_{2,1} = dx_5 \\
\theta_{2,2} &= dx_4 - x_3dx_3.
\end{aligned}$$

It is easy to see that the above obtained 1-forms are linearly independent. Hence, according to (8.5), we get

$$\Delta = \operatorname{span}\left\{\theta_{1,1}, \theta_{1,2}, \theta_{1,3}, \theta_{2,1}, \theta_{2,2}\right\},$$

which is of rank 5. Based on Definition 8.1, we obtain the related observability indices as $r_1 = 3$ and $r_2 = 2$. Moreover, according to (8.26) we get

$$\Delta^{\perp} = \operatorname{span}\left\{\tau_6\right\},$$

where

$$\tau_6 = \frac{\partial}{\partial x_6}.$$

Based on the proposed method, we solve (8.7) to get one following solution

$$\bar{\tau}_{1,1} = \frac{\partial}{\partial x_1} \text{ and } \bar{\tau}_{2,1} = \frac{\partial}{\partial x_4}$$

based on which we can calculate the other vector fields as follows:

$$\begin{aligned}\overline{\tau}_{1,2} &= \tfrac{\partial}{\partial x_2}\\ \overline{\tau}_{1,3} &= \tfrac{\partial}{\partial x_3} + x_5\tfrac{\partial}{\partial x_1} - 2x_5\tfrac{\partial}{\partial x_4} + x_5\tfrac{\partial}{\partial x_6}\\ \overline{\tau}_{2,2} &= \tfrac{\partial}{\partial x_5} + x_3\tfrac{\partial}{\partial x_1} - 2x_3\tfrac{\partial}{\partial x_4}.\end{aligned}$$

Then we can check that $[\bar{\tau}_{i,j}, \bar{\tau}_{k,s}] = 0$ except

$$\left[\bar{\tau}_{1,3}, \bar{\tau}_{2,2}\right] = \frac{\partial}{\partial x_6} \in \Delta^{\perp}.$$

However, if we choose the following vector fields

$$\tau_{i,j} = \bar{\tau}_{i,j} \text{ for } j \neq 3$$

$$\tau_{1,3} = \bar{\tau}_{1,3} - x_5\tau_6,$$

then all vector fields commute, i.e., $[\tau_{i,j}, \tau_{k,l}] = 0$, which enables us to compute the change of coordinates.

For this, let us firstly calculate

$$\Lambda = \begin{pmatrix} 0\,0 & 1 & 0 & 0 \\ 0\,1 & 0 & 0 & 0 \\ 1\,0 & x_5 & 0 & x_3 \\ 0\,0 & 0 & 0 & 1 \\ 0\,0 & -2x_5 - 2x_3 & 1 & -2x_3 \end{pmatrix}$$

based on which we get

$$\begin{aligned} dz_{1,1} &= \omega_{1,1} = -x_5\theta_{1,1} + \theta_{1,3} - x_3 dx_5\\ &= d\left(x_1 - x_3x_5\right)\\ dz_{1,2} &= \omega_{1,2} = dx_2\\ dz_{1,3} &= \omega_{1,3} = dx_3\\ dz_{2,1} &= \omega_{2,1} = \left(2x_3 + 2x_5\right)\theta_{1,1} + 2x_3\theta_{2,1} + \theta_{2,2}\\ &= d(x_4 + 2x_3x_5)\\ z_{2,2} &= \omega_{2,2} = dx_5.\end{aligned}$$

Moreover, as we have chosen $\tau_6 = \frac{\partial}{\partial x_6}$, then for the complementary basis, we might take

$$\eta = x_6.$$

Finally, we obtain the following diffeomorphism:

$$\begin{aligned} z_{1,1} &= x_1 - x_3 x_5 \\ z_{1,2} &= x_2,\, z_{1,3} = x_3 \\ z_{2,1} &= x_4 + 2x_3 x_5 \\ z_{2,2} &= x_5,\, \eta = x_6, \end{aligned}$$

with which the studied system can be transformed into the following normal form:

$$\begin{aligned} \dot{z}_{1,1} &= y_1^2 + y_1 y_2 \\ \dot{z}_{1,2} &= z_{1,1} + y_1 y_2 \\ z_{1,3} &= z_{1,2},\, \dot{z}_{2,1} = 2y_1^3 \\ \dot{z}_{2,2} &= z_{2,1} - 2y_1 y_2 - y_1^2 \\ \dot{\eta} &= -\eta + z_{1,2} z_{2,2} \\ y_1 &= z_{1,3},\, y_2 = z_{2,2}. \end{aligned}$$

8.7 Observer Design

8.7.1 Reduced-Order Luenberger-Like Observer

The concept of reduced-order observer was firstly introduced in [7], which needs only estimate unmeasurable states of the studied system. It has been generalized for nonlinear dynamical systems by imposing the Lipschitz conditions for nonlinear terms [9, 11] and invariant manifold [5]. Besides, such a technique has been used to design observer for descriptor system [4] and chaotic synchronization system [12].

To design an observer for the normal form (8.2) or (8.22)–(8.24), we consider in this chapter a reduced-ordered observer. For this, let us decompose the observable state z of the normal form (8.2) or (8.22)–(8.24) by applying the following linear transformation

$$\begin{bmatrix} \underline{z} \\ \bar{z} \end{bmatrix} = \begin{bmatrix} \underline{z} \\ \bar{y} \end{bmatrix} = Tz,$$

with which (8.2) or (8.22)–(8.24) can be then decomposed as

$$\begin{aligned} \dot{\underline{z}} &= A_{11}\underline{z} + A_{12}\bar{z} + \bar{\beta}_1(\bar{y}) \\ \dot{\bar{z}} &= A_{21}\underline{z} + A_{22}\bar{z} + \bar{\beta}_2(\bar{y}) \\ \bar{y} &= \bar{z}, \end{aligned} \tag{8.38}$$

where $\underline{z} \in \mathbb{R}^{n-m}$ and $\bar{z} \in \mathbb{R}^m$, and

$$\begin{bmatrix} A_{11} & A_{12} \\ A_{21} & A_{22} \end{bmatrix} = TA_O T^{-1}, \quad [0, I] = C_O T^{-1}, \quad \begin{bmatrix} \bar{\beta}_1(\bar{y}) \\ \bar{\beta}_2(\bar{y}) \end{bmatrix} = T\beta(\bar{y}).$$

Since the pair (A_O, C_O) in (8.2) or (8.22)–(8.24) is observable, i.e.,

$$\text{rank}\begin{bmatrix}\lambda I - A_O \\ C_O\end{bmatrix} = n, \forall \lambda \in \mathbb{C}.$$

Due to the fact that T is invertible, thus

$$\begin{aligned}\text{rank}\begin{bmatrix}\lambda I - A_O \\ C_O\end{bmatrix} &= \text{rank}\begin{bmatrix}\lambda I - TA_OT^{-1} \\ C_OT^{-1}\end{bmatrix} \\ &= \text{rank}\begin{bmatrix}\lambda I - A_{11} & -A_{12} \\ -A_{21} & \lambda I - A_{22} \\ 0 & I\end{bmatrix} \\ &= \text{rank}\begin{bmatrix}\lambda I - A_{11} & 0 \\ -A_{21} & 0 \\ 0 & I\end{bmatrix} \\ &= m + \text{rank}\begin{bmatrix}\lambda I - A_{11} \\ A_{21}\end{bmatrix}.\end{aligned}$$

Therefore, we obtain

$$\text{rank}\begin{bmatrix}\lambda I - A_{11} \\ A_{21}\end{bmatrix} = n - m, \forall \lambda \in \mathbb{C},$$

which implies that the observability of (A_O, C_O) ensures that (A_{11}, A_{21}) in (8.38) is observable as well. Consequently, there exists a vector K such that $(A_{11} - KA_{21})$ is Hurwitz.

Consider now the following dynamics:

$$\begin{aligned}\dot{\zeta} &= (A_{11} - KA_{21})\zeta + \bar{\beta}_1(\bar{y}) + (A_{11}K + A_{12} - KA_{22} - KA_{21}K)\bar{y} - K\bar{\beta}_2(\bar{y}) \\ \hat{\underline{z}} &= \zeta + K\bar{y}.\end{aligned} \tag{8.39}$$

Note $e = \underline{z} - \hat{\underline{z}}$, then according to (8.38) and (8.39), we obtain

$$\begin{aligned}\dot{e} &= \dot{\underline{z}} - \dot{\hat{\underline{z}}} \\ &= A_{11}\underline{z} + A_{12}\bar{z} + \bar{\beta}_1(\bar{y}) - \dot{\zeta} - K\dot{\bar{z}} \\ &= A_{11}\underline{z} + (A_{11} - KA_{21})\hat{\underline{z}} - KA_{21}\underline{z} \\ &= (A_{11} - KA_{21})e.\end{aligned}$$

Due to the fact that there exists a K such that $(A_{11} - KA_{21})$ is Hurwitz, therefore, e is asymptotically stable, and we have

$$\lim_{t\to+\infty} \hat{\underline{z}}(t) = \underline{z}(t).$$

In summary, the dynamics (8.39) is an asymptotic reduced-order observer of (8.2) or (8.22)–(8.24).

8.7.2 Design Procedure

The following summarizes the complete procedure when applying the proposed method for a given multi-output nonlinear dynamical system of the form (8.1):

Step 1: Determine the observability indices r_i for $1 \leq i \leq m$ and compute the associated observability 1-forms $\theta_{i,j}$ for $1 \leq i \leq m$ and $1 \leq j \leq r_i$;

Step 2: Determine Δ, $\Delta^{\perp}$ and seek a commutative basis $\{\bar{\tau}_{r+1}, \ldots, \bar{\tau}_n\}$ that spans $\Delta^{\perp}$;

Step 3: Choose any solution $\tau_{i,1}$ of (8.7) for $1 \leq i \leq m$, and eliminate all its terms modulo $\Delta^{\perp}$ which yields $\bar{\tau}_{i,1}$;

Step 4: Iteratively, for $2 \leq j \leq r_i$, compute $[\bar{\tau}_{i,j}, f]$ and eliminate all its terms modulo $\Delta^{\perp}$ which yields $\bar{\tau}_{i,j}$;

Step 5: Calculate the multi-valued matrix Λ and the associated r differential 1-forms ω;

Step 6: Integrating ω yields $z = \phi_1(x)$, and then we can choose $\eta = \phi_2(x)$, complementary to $\phi_1(x)$ such that $\phi(x) = [\phi_1^T(x), \phi_2^T(x)]^T$ forms a change of coordinates;

Step 7: For (8.1), applying $\phi(x) = (z^T, \eta^T)^T$ to obtain the normal form (8.22)–(8.24), and then design the proposed reduced-order Luenberger-like observer (8.39) to estimate z of (8.2) or (8.22)–(8.24).

References

1. Bodson, M., Chiasson, J., Novotnak, R.: High-performance induction motor control via input-output linearization. IEEE Control Syst. Mag. **14**(4), 25–33 (1994)
2. Boutat, D., Liu, D.Y.: Observer design for a class of non-linear systems with linearisable error dynamics. IET Control Theory Appl. **9**(15), 2298–2304 (2015)
3. Boutat, D., Zheng, G., Barbot, J.P., Hammouri, H.: Observer error linearization multi-output depending. In: Proceedings of the 45th IEEE Conference on Decision and Control, pp. 5394–5399 (2006)
4. Darouach, M., Zasadzinski, M., Hayar, M.: Reduced-order observer design for descriptor systems with unknown inputs. IEEE Trans. Autom. Control **41**(7), 1068–1072 (1996)
5. Karagiannis, D., Carnevale, D., Astolfi, A.: Invariant manifold based reduced-order observer design for nonlinear system. IEEE Trans. Autom. Control **53**(11), 2602–2614 (2008)
6. Krener, A., Respondek, W.: Nonlinear observers with linearizable error dynamics. SIAM J. Control Optim. **23**(2), 197–216 (1985)
7. Luenberger, D.: An introduction to observers. IEEE Trans. Autom. Control **16**(6), 596–602 (1971)
8. Saadi, W., Boutat, D., Zheng, G., Sbita, L., Yu, L.: Algorithm to compute nonlinear partial observer normal form with multiple outputs. IEEE Trans. Autom. Control **65**(6), 2700–2707 (2020)

9. Sundarpandian, V.: Reduced order observer design for nonlinear systems. Appl. Math. Lett. **19**(9), 936–941 (2006)
10. Xia, X.H., Gao, W.B.: Nonlinear observer design by observer error linearization. SIAM J. Control Optim. **27**(1), 199–216 (1989)
11. Xu, M.: Reduced-order observer design for one-sided Lipschitz nonlinear systems. IMA J. Math. Control Inf. **26**(3), 299–317 (2009)
12. Zheng, G., Boutat, D.: Synchronisation of chaotic systems via reduced observers. IET Control Theory Appl. **5**(2), 308–314 (2011)
13. Zheng, G., Boutat, D., Barbot, J.P.: Multi-output dependent observability normal form. Nonlinear Anal.: Theory Methods Appl. **70**(1), 404–418 (2009)

Chapter 9
Extension to Nonlinear Singular Dynamical Systems

Abstract This chapter aims at extending the differential geometric method to design observers for nonlinear singular dynamical systems. Singular systems widely exist in engineering systems, such as chemical system, biological system, electrical circuit and so on. These systems are governed by mixing differential and algebraic equations, which is the key difference with respect to regular systems [5, 6]. Due to this reason, many well-defined concepts relative to observation problem for regular (non-singular) systems have to be reconsidered for singular ones. For such systems, observability has been analyzed in [3, 10], and different types of observers have been proposed in [4, 9, 11]. In this chapter, we will show how to apply differential geometric method to design observers for such a nonlinear singular system [12]. The basic idea is to regularize the studied singular system into a nonlinear regular system with the injection of the output derivative, then seek a diffeomorphism to transform the regularized system into an observer normal form with output derivative injection, based on which a Luenberger-like observer is designed.

9.1 Problem Statement

Consider the following multi-output nonlinear singular dynamical system

$$\begin{aligned} E\dot{x} &= f(x) \\ y &= h(x), \end{aligned} \tag{9.1}$$

where $x \in \mathscr{X} \subseteq \mathbb{R}^n$ is the state vector, $\mathscr{X}$ is an open domain of $\mathbb{R}^n$, $f : \mathscr{X} \to \mathbb{R}^n$ and $h = [h_1, \ldots, h_m]^T : \mathscr{X} \to \mathbb{R}^m$ are both assumed to be smooth, with $E \in \mathbb{R}^{n\times n}$ being singular, i.e., rank$E < n$. Without loss of generality, it is assumed that $0 \in \mathscr{X}$, $f(0) = 0$ and $h(0) = 0$.

Note that if the matrix E has full row rank, i.e., rank$E = n$, then system (9.1) is equivalent to

$$\begin{aligned} \dot{x} &= E^{-1} f(x) \\ y &= h(x), \end{aligned}$$

D. Boutat and G. Zheng, *Observer Design for Nonlinear Dynamical Systems*,
Lecture Notes in Control and Information Sciences 487,
https://doi.org/10.1007/978-3-030-73742-9_9

which is of the general form (8.1), and we can apply the results developed in Chap. 8 to seek the desired change of coordinates.

Note that singular system contains not only differential equations, but also algebraic ones [1, 2]. Since differential geometric approach has been developed for dynamical system governed by an ordinary differential equation, we are wondering whether it is still possible to apply this approach to seek a diffeomorphism for the singular systems, with which system (9.1) can be transformed into a simpler observer normal form? In order to answer this natural question, the following example presents an idea of how to treat the singular case.

Example 9.1 Consider the following nonlinear singular system:

$$\begin{bmatrix} 0 & 1 & 0 \\ 0 & 0 & 1 \\ 0 & 0 & 0 \end{bmatrix} \begin{bmatrix} \dot{x}_1 \\ \dot{x}_2 \\ \dot{x}_3 \end{bmatrix} = \begin{bmatrix} x_1 - \frac{3}{2}x_3^2 + \frac{3}{2}x_3^{\frac{1}{2}}x_2 \\ x_2 - x_3^{\frac{3}{2}} \\ x_3 + x_1 \end{bmatrix}. \qquad (9.2)$$
$$y = x_1$$

It is clear that $E = \begin{bmatrix} 0 & 1 & 0 \\ 0 & 0 & 1 \\ 0 & 0 & 0 \end{bmatrix}$ is singular, and this explains as well that the dynamics $\dot{x}_1$ is missing.

By applying the following change of coordinates,

$$\begin{bmatrix} z_1 \\ z_2 \\ z_3 \end{bmatrix} = \begin{bmatrix} x_2 - x_3^{\frac{3}{2}} \\ x_3 \\ x_1 \end{bmatrix},$$

it can be checked that the studied system (9.2) can be transformed into the following form:

$$\begin{aligned} \dot{z}_1 &= y \\ \dot{z}_2 &= z_1 \\ \dot{z}_3 &= z_2 + y + \dot{y} \\ y &= z_3. \end{aligned} \qquad (9.3)$$

In fact, the above form can be written in a more compact one as

$$\begin{aligned} \dot{z} &= A_O z + \beta(y, \dot{y}) \\ y &= C_O z, \end{aligned}$$

which can be seen as a generalization of the observer normal form with output injection (3.3). Since this general form contains both y and $\dot{y}$, therefore we call it the observer normal form with output derivative injection.

This example reveals at least two important facts:

(1) Even no dynamics has been defined for x_1 by a differential equation, if certain properties are satisfied, it is possible to regularize this singular system into a regular one. For example, in (9.3), the third equation is in fact an algebraic equation, but it has been regularized by using $\dot{y}$;
(2) Due to the effect of regularization, the transformed normal form will contain both y and $\dot{y}$.

□

From the above analysis, for the nonlinear singular system (9.1), this chapter investigates how to seek a diffeomorphism $z = \phi(x)$ with $z = (z_1^T, z_2^T, \ldots, z_{\bar{m}}^T)^T$ and $z_i = \left(z_{i,1}, z_{i,2}, \ldots, z_{i,r_i}\right)^T$, for $1 \leq i \leq \bar{m}$ with $\sum_{i=1}^{\bar{m}} r_i = n$, via which system (9.1) can be transformed into the following observer normal form with output derivative injection

$$\begin{aligned} \dot{z}_i &= A_{O_i} z_i + \alpha_i(\bar{y}) + \beta_i(\bar{y})\dot{\bar{y}}, \text{ for } 1 \leq i \leq \bar{m} \\ \bar{y}_i &= C_{O_i} z_i = z_{i,r_i} \end{aligned} \tag{9.4}$$

with A_{O_i} and C_{O_i} being of Brunovsky form

$$A_{O_i} = \begin{pmatrix} 0 & \cdots & 0 & 0 & 0 \\ 1 & \cdots & 0 & 0 & 0 \\ \vdots & \ddots & \ddots & \cdots & \vdots \\ 0 & \cdots & 1 & 0 & 0 \\ 0 & \cdots & 0 & 1 & 0 \end{pmatrix}, C_{O_i} = (0, \ldots, 0, 1)$$

and α_i and β_i being, respectively, r_i and $r_i \times \bar{m}$ matrix.

9.2 Transformation into Regular System

Since the objective of this book focuses on the observer design by applying differential geometric method, before presenting the procedure to transform nonlinear singular system (9.1) into a regular one, it is worth understanding what is the condition to design an observer for linear singular system.

For this, we consider the following linear singular system without input

$$\begin{aligned} E\dot{x} &= \bar{A}x \\ y &= Cx \end{aligned} \tag{9.5}$$

with $x \in \mathbb{R}^n$, rank$E < n$. In [8], it has been stated that there exists a simple Luenberger-like observer

$$\begin{aligned} \dot{\xi} &= N\xi + Ly \\ \hat{x} &= \xi + Ky \end{aligned} \tag{9.6}$$

with properly chosen constant matrices N, K and L, if the following two conditions:

$$\text{rank}\begin{bmatrix} E & A \\ 0 & E \\ 0 & C \end{bmatrix} = n + \text{ rank } E \tag{9.7}$$

and

$$\text{rank}\begin{bmatrix} sE - A \\ C \end{bmatrix} = n, \forall s \in \mathbb{C}, \text{Re}(s) \geq 0 \tag{9.8}$$

are satisfied.

For the purpose of generalizing the above two conditions for nonlinear singular system (9.1), we can impose the following similar condition.

Assumption 9.1 It is assumed that, for E, $f(x)$ and $h(x)$ defined in (9.1), the following condition:

$$\text{rank}\begin{bmatrix} E & \frac{\partial f(x)}{\partial x} \\ 0 & E \\ 0 & \frac{\partial h(x)}{\partial x} \end{bmatrix} = n + \text{rank } E \tag{9.9}$$

is satisfied.

It is worth highlighting that (9.9) in Assumption 9.1 is an extension of (9.7) to deal with nonlinear singular case. Indeed, if we set $f(x) = Ax$ and $h(x) = Cx$, then (9.9) is equivalent to (9.7), since

$$\text{rank}\begin{bmatrix} E & \frac{\partial f(x)}{\partial x} \\ 0 & E \\ 0 & \frac{\partial h(x)}{\partial x} \end{bmatrix} = \text{rank}\begin{bmatrix} E & A \\ 0 & E \\ 0 & C \end{bmatrix} = n + \text{rank } E,$$

which has been stated as the sufficient and necessary observability condition for linear singular systems.

Note rank $E = q < n$, therefore we can always find two elementary matrices with appropriate dimension, noted as T and S, such that

$$SET = \begin{bmatrix} I_q & 0 \\ 0 & 0 \end{bmatrix}. \tag{9.10}$$

Hence, by introducing the new change of variables $\zeta = T^{-1}x$, (9.1) can be written as

$$\begin{aligned} SET\dot{\zeta} &= Sf(T\zeta) \\ y &= h(T\zeta) \end{aligned}$$

or equivalently

$$\begin{aligned} \bar{E}\dot{\zeta} &= \tilde{f}(\zeta) \\ y &= \tilde{h}(\zeta), \end{aligned} \tag{9.11}$$

where $\bar{E} = \begin{bmatrix} I_q & 0 \\ 0 & 0 \end{bmatrix}$, $\tilde{f}(\zeta) = Sf(T\zeta)$ and $\tilde{h}(\zeta) = h(T\zeta)$. Then the above system can be decomposed into the following two subsystems:

$$\begin{aligned} \dot{\zeta}_1 &= \tilde{f}_1(\zeta_1, \zeta_2) \\ 0 &= \tilde{f}_2(\zeta_1, \zeta_2) \\ y &= \tilde{h}(\zeta_1, \zeta_2), \end{aligned} \tag{9.12}$$

with $\zeta = \begin{bmatrix} \zeta_1 \\ \zeta_2 \end{bmatrix}$, $\zeta_1 \in \mathbb{R}^q$, $\zeta_2 \in \mathbb{R}^{n-q}$, $\tilde{f} = \begin{bmatrix} \tilde{f}_1 \\ \tilde{f}_2 \end{bmatrix}$, and we have the following result.

Lemma 9.1 *The following condition*

$$\operatorname{rank} \begin{bmatrix} \frac{\partial \tilde{f}_2(\zeta)}{\partial \zeta_2} \\ \frac{\partial \tilde{h}(\zeta)}{\partial \zeta_2} \end{bmatrix} = n - q \tag{9.13}$$

is true if and only if (9.9) is satisfied.

Proof Due to the fact that S and T in (9.10) are both invertible, therefore, we obtain

$$\begin{aligned} \operatorname{rank} \begin{bmatrix} E & \frac{\partial f(x)}{\partial x} \\ 0 & E \\ 0 & \frac{\partial h(x)}{\partial x} \end{bmatrix} &= \operatorname{rank} \begin{bmatrix} S & 0 & 0 \\ 0 & S & 0 \\ 0 & 0 & I_m \end{bmatrix} \begin{bmatrix} E & \frac{\partial f(x)}{\partial x} \\ 0 & E \\ 0 & \frac{\partial h(x)}{\partial x} \end{bmatrix} \begin{bmatrix} T & 0 \\ 0 & T \end{bmatrix} \\ &= \operatorname{rank} \begin{bmatrix} SET & S\frac{\partial f(x)}{\partial x}T \\ 0 & SET \\ 0 & \frac{\partial h(x)}{\partial x}T \end{bmatrix}. \end{aligned}$$

According to the definition $x = T\zeta$, we get

$$S\frac{\partial f(x)}{\partial x}T = \frac{\partial \tilde{f}(\zeta)}{\partial \zeta}$$

and

$$\frac{\partial h(x)}{\partial x}T = \frac{\partial \tilde{h}(\zeta)}{\partial \zeta}.$$

Hence, we have

$$\begin{aligned}\operatorname{rank}\begin{bmatrix} E & \frac{\partial f(x)}{\partial x} \\ 0 & E \\ 0 & \frac{\partial h(x)}{\partial x}\end{bmatrix} &= \operatorname{rank}\begin{bmatrix} SET & \frac{\partial \tilde{f}(\zeta)}{\partial \zeta} \\ 0 & SET \\ 0 & \frac{\partial \tilde{h}(\zeta)}{\partial \zeta}\end{bmatrix} \\ &= \operatorname{rank}\begin{bmatrix} I_q & 0 & \frac{\partial \tilde{f}_1(\zeta)}{\partial \zeta_1} & \frac{\partial \tilde{f}_1(\zeta)}{\partial \zeta_2} \\ 0 & 0 & \frac{\partial \tilde{f}_2(\zeta)}{\partial \zeta_1} & \frac{\partial \tilde{f}_2(\zeta)}{\partial \zeta_2} \\ 0 & 0 & I_q & 0 \\ 0 & 0 & 0 & 0 \\ 0 & 0 & \frac{\partial \tilde{h}(\zeta)}{\partial \zeta_1} & \frac{\partial \tilde{h}(\zeta)}{\partial \zeta_2}\end{bmatrix} \\ &= 2q + \operatorname{rank}\begin{bmatrix} \frac{\partial \tilde{f}_2(\zeta)}{\partial \zeta_2} \\ \frac{\partial \tilde{h}(\zeta)}{\partial \zeta_2}\end{bmatrix}.\end{aligned}$$

Consequently, the condition (9.9) is equivalent to rank $\begin{bmatrix} \frac{\partial \tilde{f}_2(\zeta)}{\partial \zeta_2} \\ \frac{\partial \tilde{h}(\zeta)}{\partial \zeta_2}\end{bmatrix} = n - q.$ □

In the following, we show how to transform the studied nonlinear singular system (9.1) into the associated nonlinear regular one and such a procedure is called regularization.

It is clear that if rank $\begin{bmatrix} \frac{\partial \tilde{f}_2(\zeta)}{\partial \zeta_2} \\ \frac{\partial \tilde{h}(\zeta)}{\partial \zeta_2}\end{bmatrix} = n - q$, then we can find an invertible matrix $Q(\zeta) = \begin{bmatrix} Q_1(\zeta) \\ Q_2(\zeta)\end{bmatrix} \in \mathbb{R}^{(m+n-q)\times(m+n-q)}$ which yields

$$Q(\zeta)\begin{bmatrix} \frac{\partial \tilde{f}_2(\zeta)}{\partial \zeta_2} \\ \frac{\partial \tilde{h}(\zeta)}{\partial \zeta_2}\end{bmatrix} = \begin{bmatrix} I_{n-q} \\ 0_m \end{bmatrix} \tag{9.14}$$

and $Q_1(\zeta)\begin{bmatrix} \frac{\partial \tilde{f}_2(\zeta)}{\partial \zeta_2} \\ \frac{\partial \tilde{h}(\zeta)}{\partial \zeta_2}\end{bmatrix} = I_{n-q}$. Now let us calculate the derivative of the algebraic equation in (9.12), which gives

$$\begin{bmatrix} 0 \\ \dot{y}\end{bmatrix} = \begin{bmatrix} \frac{\partial \tilde{f}_2(\zeta)}{\partial \zeta} \\ \frac{\partial \tilde{h}(\zeta)}{\partial \zeta}\end{bmatrix}\dot{\zeta} = \begin{bmatrix} \frac{\partial \tilde{f}_2(\zeta)}{\partial \zeta_1} \\ \frac{\partial \tilde{h}(\zeta)}{\partial \zeta_1}\end{bmatrix}\tilde{f}_1(\zeta) + \begin{bmatrix} \frac{\partial \tilde{f}_2(\zeta)}{\partial \zeta_2} \\ \frac{\partial \tilde{h}(\zeta)}{\partial \zeta_2}\end{bmatrix}\dot{\zeta}_2.$$

For the above equation, by applying (9.14), we can obtain

$$\dot{\zeta}_2 = -Q_1(\zeta)\begin{bmatrix} \frac{\partial \tilde{f}_2(\zeta)}{\partial \zeta_1} \\ \frac{\partial \tilde{h}(\zeta)}{\partial \zeta_1}\end{bmatrix}\tilde{f}_1(\zeta) + Q_1(\zeta)\begin{bmatrix} 0 \\ \dot{y}\end{bmatrix}.$$

This implies that we can regularize the algebraic equation in (9.12) by adding the obtained artificial dynamics, and this leads to the following regular system with output derivatives

$$\begin{cases} \dot{\zeta} = \bar{f}(\zeta) + g(\zeta)\dot{y} \\ \bar{y} = \bar{h}(\zeta) \end{cases} \tag{9.15}$$

with

$$\bar{f}(x) = \begin{bmatrix} \tilde{f}_1(\zeta) \\ -Q_1(\zeta) \begin{bmatrix} \frac{\partial \tilde{f}_2(\zeta)}{\partial \zeta_1} \\ \frac{\partial \tilde{h}(\zeta)}{\partial \zeta_1} \end{bmatrix} \tilde{f}_1(\zeta) \end{bmatrix} \tag{9.16}$$

$$g(\zeta) = \begin{bmatrix} 0 \\ Q_1(\zeta) \end{bmatrix} \tag{9.17}$$

and

$$\bar{y} = \begin{bmatrix} 0 \\ y \end{bmatrix}, \bar{h}(\zeta) = \begin{bmatrix} \tilde{f}_2(\zeta) \\ \tilde{h}(\zeta) \end{bmatrix}. \tag{9.18}$$

We would like to highlight that $\bar{y}$ (which can be regarded as a new output) contains two parts: measurable variable y and a deduced algebraic equation from (9.1).

With the obtained new output $\bar{y}$, for the sake of simplicity and without loss of generality, based on the regularized form (9.15), we can assume that the first $\bar{m}$ outputs are linearly independent with $\bar{m} \leq n + m - q$, i.e., $\{d\bar{h}_1, \ldots, d\bar{h}_{\bar{m}}\}$ are linearly independent. If this assumption is not satisfied, we can then perform certain algebraic manipulations to remove the dependent outputs and reorder them. To investigate the observability of the regularized system, the following assumption is needed.

Assumption 9.2 For the regularized system (9.15), it is assumed that there exist r_i for $1 \leq i \leq \bar{m}$ such that $\sum_{i=1}^{\bar{m}} r_i = n$ and

$$\text{rank} \begin{bmatrix} d\bar{h}_1 \\ \vdots \\ dL_{\bar{f}}^{r_i - 1}\bar{h}_1 \\ \vdots \\ d\bar{h}_{\bar{m}} \\ \vdots \\ dL_{\bar{f}}^{r_{\bar{m}} - 1}\bar{h}_{\bar{m}} \end{bmatrix} = n, \tag{9.19}$$

where $L_{\bar{f}}^k\bar{h}_i$ for $1 \leq k \leq r_i$.

Clearly, (9.19) implies that the regularized system is locally observable. Moreover, we would like to show that the condition (9.19) is a natural generalization of the condition (9.8) when studying nonlinear singular systems.

Proposition 9.1 *Consider system (9.1) with $f(x) = Ax$ and $h(x) = Cx$. If Assumption 9.1 is fulfilled, then (9.19) implies (9.8).*

Proof For system (9.1) with $f(x) = Ax$ and $h(x) = Cx$, we can always find two invertible matrices T and S such that (9.1) might be transformed via the diffeomorphism $x = T\zeta$ into the following system:

$$\begin{aligned} \bar{E}\dot{\zeta} &= \tilde{A}\zeta \\ y &= \tilde{C}\zeta, \end{aligned} \tag{9.20}$$

where $\bar{E} = SET = \begin{bmatrix} I_q & 0 \\ 0 & 0 \end{bmatrix}$, $\tilde{A} = SAT$ and $\tilde{C} = CT$. The sake of simplicity denote $\bar{E}_q = [I_q, 0]$, $\tilde{A} = \begin{bmatrix} \tilde{A}_1 \\ \tilde{A}_2 \end{bmatrix}$ and $\bar{C} = \begin{bmatrix} \tilde{A}_2 \\ \tilde{C} \end{bmatrix}$, then (9.20) is equivalent to

$$\begin{aligned} \bar{E}_q\dot{\zeta} &= \dot{\zeta}_1 = \tilde{A}_1\zeta \\ \bar{y} &= \bar{C}\zeta, \end{aligned} \tag{9.21}$$

where $\zeta_1 \in \mathbb{R}^q$ and $\bar{y} = \begin{bmatrix} 0 \\ y \end{bmatrix}$.

Now let us rewrite $\bar{C}$ as

$$\bar{C} = \left[\bar{C}_1, \bar{C}_2\right].$$

Then, based on the result stated in Lemma 9.1, if Assumption 9.1 is fulfilled, we can obtain

$$\text{rank}\,\bar{C}_2 = n - q.$$

Therefore, we can find an invertible matrix Q_1 such that $Q_1\bar{C}_2 = I_{n-q}$.

Besides, due to the fact that

$$\text{rank}\begin{bmatrix} \bar{E}_q \\ \bar{C} \end{bmatrix} = \text{rank}\begin{bmatrix} I_q & 0 \\ \bar{C}_1 & \bar{C}_2 \end{bmatrix} = n,$$

then there exists an invertible matrix $P = \begin{bmatrix} I_q & 0 \\ -Q_1\bar{C}_1 & Q_1 \end{bmatrix}$ such that

$$P\begin{bmatrix} \bar{E}_q \\ \bar{C} \end{bmatrix} = I_n.$$

For system (9.21), let us consider its output derivative which can be written as

$$\begin{aligned} \bar{E}_q\dot{\zeta} &= \tilde{A}_1\zeta \\ \bar{C}\dot{\zeta} &= \dot{\bar{y}}. \end{aligned}$$

Note that we have $P\begin{bmatrix} \bar{E}_q \\ \bar{C} \end{bmatrix} = I_n$, then the above dynamics can be transformed, by multiplying both side the invertible matrix P, into the following normal form with output derivative injection

$$\begin{aligned} \dot{\zeta} &= \bar{A}\zeta + P\begin{bmatrix} 0 \\ \dot{y} \end{bmatrix} \\ \bar{y} &= \bar{C}\zeta, \end{aligned}$$

where

$$\bar{A} = \begin{bmatrix} \tilde{A}_1 \\ -Q_1\bar{C}_1\tilde{A}_1 \end{bmatrix}.$$

Therefore, for system (9.1) with $f(x) = Ax$ and $h(x) = Cx$, the condition (9.19) is indeed equivalent to the observability condition for linear time-invariant dynamical systems, i.e., the pair $(\bar{A}, \bar{C})$ is observable. According to the results stated in Chap. 1, this is equivalent to the following condition:

$$\operatorname{rank}\begin{bmatrix} sI_n - \bar{A} \\ \bar{C} \end{bmatrix} = n, \forall s \in \mathbb{C}.$$

Consequently, we obtain

$$\begin{aligned} \operatorname{rank}\begin{bmatrix} sI_n - \bar{A} \\ \bar{C} \end{bmatrix} &= \operatorname{rank}\begin{bmatrix} s\bar{E}_q - \tilde{A}_1 \\ s\,[0, I_{n-q}] + Q_1\bar{C}_1\tilde{A}_1 \\ \bar{C} \end{bmatrix} = \operatorname{rank}\begin{bmatrix} s\bar{E}_q - \tilde{A}_1 \\ s\,[0, Q_1\bar{C}_2] + Q_1\bar{C}_1\tilde{A}_1 \\ \bar{C} \end{bmatrix} \\ &= \operatorname{rank}\begin{bmatrix} s\bar{E}_q - \tilde{A}_1 \\ sQ_1\bar{C} - sQ_1\bar{C}_1\bar{E}_q + Q_1\bar{C}_1\tilde{A}_1 \\ \bar{C} \end{bmatrix} \\ &= \operatorname{rank}\begin{bmatrix} I_q & 0 & 0 \\ -Q_1\bar{C}_1 & Q_1 & 0 \\ 0 & 0 & I_{n-q+m} \end{bmatrix}\begin{bmatrix} s\bar{E}_q - \tilde{A}_1 \\ s\bar{C} \\ \bar{C} \end{bmatrix} \\ &= \operatorname{rank}\begin{bmatrix} s\bar{E}_q - \tilde{A}_1 \\ \bar{C} \end{bmatrix} = \operatorname{rank}\begin{bmatrix} s\bar{E}_q - \tilde{A}_1 \\ -\tilde{A}_2 \\ \tilde{C} \end{bmatrix} = \operatorname{rank}\begin{bmatrix} s\bar{E} - \tilde{A} \\ \tilde{C} \end{bmatrix} \\ &= n. \end{aligned}$$

Since $\bar{E} = SET = \begin{bmatrix} I_q & 0 \\ 0 & 0 \end{bmatrix}$, $\tilde{A} = SAT$ and $\tilde{C} = CT$, then we get

$$\operatorname{rank}\begin{bmatrix} s\bar{E} - \tilde{A} \\ \tilde{C} \end{bmatrix} = \operatorname{rank}\begin{bmatrix} sE - A \\ C \end{bmatrix} = n$$

for all $s \in \mathbb{C}$. Finally, we have proved that, for system (9.1) with $f(x) = Ax$ and $h(x) = Cx$, if Assumption 9.1 is fulfilled, then (9.19) implies (9.8). □

9.3 Necessary and Sufficient Conditions

For the regularized system, this section gives necessary and sufficient condition to transform it into the output derivative injection form (9.4).

Recall that if Assumptions 9.1 and 9.2 for nonlinear singular dynamical system (9.1) are both fulfilled, then this singular system might be regularized into

$$\begin{aligned} \dot{\zeta} &= \bar{f}(\zeta) + g(\zeta)\dot{y} \\ \bar{y} &= \bar{h}(\zeta), \end{aligned} \tag{9.22}$$

where $\bar{f}$, g and $\bar{h}$ were given in (9.16)–(9.18), respectively, where the output components $\{d\bar{h}_1, \ldots, d\bar{h}_{\bar{m}}\}$ are assumed to be linearly independent.

In order to transform the regularized system (9.22) into (9.4), by following the procedure introduced in Chap. 8, we need firstly calculate the following 1-forms

$$\theta_{i,k} = dL_{\bar{f}}^{k-1}\bar{h}_i, \text{ for } 1 \leq i \leq \bar{m} \text{ and } 1 \leq k \leq r_i,$$

which enables us to define the following distribution:

$$\Delta = \operatorname{span}\left\{\theta_{j,k}, 1 \leq j \leq \bar{m} \text{ and } 1 \leq k \leq r_j\right\}. \tag{9.23}$$

Like what we have done in Chap. 8 for nonlinear dynamical systems with multiple outputs, the family of vector fields $\left(\tau_{i,1}\right)_{1\leq i\leq \bar{m}}$ can be calculated via the following equations:

$$\begin{aligned} \theta_{i,r_i}\left(\tau_{i,1}\right) &= 1, \text{ for } 1 \leq i \leq \bar{m} \\ \theta_{i,k}\left(\tau_{i,1}\right) &= 0, \text{ for } 1 \leq k \leq r_i - 1 \\ \theta_{j,k}\left(\tau_{i,1}\right) &= 0, \text{ for } j < i \text{ and } 1 \leq k \leq r_j \\ \theta_{j,k}\left(\tau_{i,1}\right) &= 0, \text{ for } j > i \text{ and } 1 \leq k \leq r_j. \end{aligned} \tag{9.24}$$

Since Assumption 9.2 is supposed to be fulfilled for the regularized system (9.22), then the vector fields $(\tau_{i,1})_{1\leq i\leq \bar{m}}$ satisfying (9.24) can generate the following frame

$$\tau = \left(\ldots, \tau_{i,j}, \ldots\right)_{1\leq i\leq \bar{m}, 1\leq j\leq r_i}$$

with

$$\tau_{i,k} = \left[\tau_{i,k-1}, f\right], \text{ for } 1 \leq i \leq \bar{m} \text{ and } 2 \leq k \leq r_i. \tag{9.25}$$

Following the similar procedure in Chap. 8, denote the 1-forms as

$$\theta = \left(\theta_{1,1}, \ldots, \theta_{1,r_1}, \ldots, \theta_{\bar{m},1}, \ldots, \theta_{\bar{m},r_{\bar{m}}}\right)^T$$

and calculate

$$\Lambda = \theta(\tau).$$

Since the regularized system (9.22) is supposed to be observable in the sense that the observability rank condition (9.19) is satisfied, therefore Λ is invertible, which enables us to compute the following matrix:

$$\omega = \Lambda^{-1}\theta. \tag{9.26}$$

Therefore, we can extend the result stated in Theorem 8.1 of Chap. 8 to treat the regularized system (9.22).

Theorem 9.1 *For system (9.22), suppose that Assumptions 9.1 and 9.2 are fulfilled. Then there exists a local diffeomorphism $z = \phi(\zeta)$ transforming (9.22) into the output derivative injection normal form (9.4), if and only if*

(1) $[\tau_{i,j}, \tau_{s,l}] = 0$, *for* $1 \le i \le \bar{m}$, $1 \le j \le r_i$, $1 \le s \le \bar{m}$ *and* $1 \le l \le r_s$;
(2) $\dim[\Delta_i] = \dim[\Delta_i \cap \Delta]$ *for* $1 \le i \le \bar{m}$, *where* Δ *is defined in (9.23) and* Δ_i *is defined in (8.9);*
(3) $\left[\tau_{i,j}, g_k\right] = 0$ *for* $1 \le i \le \bar{m}$, $1 \le j \le r_i - 1$, $1 \le k \le \bar{m}$.

Proof In Chap. 8, it has been stated in Theorem 8.1 that the first two conditions in Theorem 9.1 are necessary and sufficient to guarantee that we can find a local diffeomorphism $z = \phi(\zeta)$ such that (9.22) without the term $g(\zeta)\dot{y}$ can be transformed into (8.2).

Therefore, it remains to prove that the third condition of Theorem 9.1 is necessary and sufficient such that the Jacobian of the deduced diffeomorphism $z = \phi(\zeta)$ can transform g_k into $\beta_k(\bar{y})$ for $1 \le k \le \bar{m}$.

For this, let us recall the results stated in Chap. 3. In Theorem 3.3, we have studied the nonlinear dynamical systems with single-input single-output where the input term $g(x)u$ plays the same role as the term $g(\zeta)\dot{y}$. For multiple inputs case, a similar result has been stated in Corollary 3.1 where the similar Lie bracket conditions (only for single-output case) have been proven to be necessary and sufficient. Therefore, the following just gives the sketch of the proof when treating multiple-output case.

Firstly, let us consider the 1-forms ω defined in (9.26), then we have

$$\omega(\tau) = I_{n\times n}.$$

The above equation means that $\omega(\tau_{i,j})$ for $1 \le i \le \bar{m}$ and $1 \le j \le r_i$ are constant. Therefore, like what we have proceeded before in the previous chapters, we can obtain

$$\begin{aligned} d\omega(\tau_{i,j}, \tau_{k,s}) &= L_{\tau_{i,j}}\omega(\tau_{k,s}) - L_{\tau_{k,s}}\omega(\tau_{i,j}) - \omega([\tau_{i,j}, \tau_{k,s}]) \\ &= -\omega([\tau_{i,j}, \tau_{k,s}]). \end{aligned}$$

The above equation implies that

$$[\tau_{i,j}, \tau_{k,s}] = 0 \text{ is equivalent to } d\omega = 0$$

since ω is an isomorphism.

Then, based on the result stated in Poincaré Lemma (see Lemma 2.1 in Chap. 2), $d\omega = 0$ implies that we can find a local diffeomorphism $z = \phi(\zeta)$ such that $\omega = d\phi$. and

$$\phi_*(\tau_{i,j}) = \frac{\partial}{\partial z_{i,j}}$$

for $1 \le i \le \bar{m}$ and $1 \le j \le r_i$. Consequently, for $1 \le k \le \bar{m}$, we obtain

$$\frac{\partial}{\partial z_{i,j}} \phi_*(g_k) = \phi_*([\tau_{i,j}, g_k]) = 0$$

if $[\tau_{i,j}, g_k] = 0$. By integration of the above equation, it is clear that

$$\phi_*(g_k) = \beta_k(\bar{y}).$$

Therefore, we proved that (9.22) can be transformed into (9.4). □

The following gives an academic example to show how to apply the above-proposed method.

Example 9.2 Consider the following nonlinear singular system

$$\begin{aligned} E\dot{x} &= f(x) \\ y &= h(x), \end{aligned} \tag{9.27}$$

where

$$E = \begin{bmatrix} -4 & 3 & 0 & 0 \\ 4 & -1 & 0 & 0 \\ -4 & 2 & 0 & 0 \\ 0 & 1 & 0 & 0 \end{bmatrix}$$

$$f(x) = \begin{bmatrix} 4x_1 - 4x_2 - 3x_4 + (3x_2 - x_3 + 2x_4)(x_3 + x_4) \\ 4x_2 - 4x_1 + x_4 + (3x_3 - x_2 + 2x_4)(x_3 + x_4) \\ 4x_3 + 6x_4 + 2(x_2 - x_3)(x_3 + x_4) \\ -x_4 + (x_2 + x_3 + 2x_4)(x_3 + x_4) \end{bmatrix}$$

and

$$h(x) = \begin{bmatrix} x_3 + x_4 + 2x_2x_4 + x_2^2 + x_4^2 \\ x_2 + x_4 \end{bmatrix}.$$

Note $\text{rank}E = q = 2 < n = 4$, then we can find the following two elementary matrices:

$$T = \begin{bmatrix} 1 & 0 & 0 & 0 \\ 1 & 1 & 0 & 0 \\ 1 & 1 & 1 & -1 \\ -1 & -1 & 0 & 1 \end{bmatrix}$$

and

$$S = \begin{bmatrix} 0 & \frac{1}{4} & 0 & \frac{1}{4} \\ \frac{1}{4} & \frac{1}{4} & \frac{1}{4} & 0 \\ 0 & \frac{1}{4} & \frac{1}{4} & -\frac{1}{4} \\ \frac{1}{4} & \frac{1}{4} & 0 & -\frac{1}{2} \end{bmatrix}$$

with which the matrix E can be transformed into

$$\bar{E} = SET = \begin{bmatrix} 1 & 0 & 0 & 0 \\ 0 & 1 & 0 & 0 \\ 0 & 0 & 0 & 0 \\ 0 & 0 & 0 & 0 \end{bmatrix}.$$

Hence, we can introduce the following change of variables:

$$\zeta = T^{-1}x = \begin{bmatrix} x_1 \\ x_2 - x_1 \\ x_3 + x_4 \\ x_2 + x_4 \end{bmatrix}$$

via which the studied nonlinear singular system can be transformed into

$$\begin{aligned} \dot{\zeta}_1 &= \zeta_2 + \zeta_3^2 \\ \dot{\zeta}_2 &= \zeta_3 + \zeta_3\zeta_4 \\ 0 &= \zeta_3 - \zeta_1 + \zeta_4 \\ \bar{y}_1 &= \zeta_3 + \zeta_4^2 \\ \bar{y}_2 &= \zeta_4. \end{aligned} \tag{9.28}$$

Besides, we can check the following condition:

$$\text{rank} \begin{bmatrix} E & \frac{\partial f(\zeta)}{\partial \zeta} \\ 0 & E \\ 0 & \frac{\partial h(\zeta)}{\partial \zeta} \end{bmatrix} = 6 = 4 + \text{rank}E,$$

which implies that Assumption 9.1 is fulfilled. Consequently, according to the result stated in Sect. 9.2, the studied singular dynamical system can be regularized into the following one:

$$\begin{aligned} \dot{\zeta} &= \begin{bmatrix} \zeta_2 + \zeta_3^2 \\ \zeta_3 + \zeta_3\zeta_4 \\ \zeta_2 + \zeta_3^2 \\ 0 \end{bmatrix} + \begin{bmatrix} 0 \\ 0 \\ -1 \\ 1 \end{bmatrix} \dot{y}_3 \\ y &= \begin{bmatrix} \zeta_3 - \zeta_1 + \zeta_4 \\ \zeta_3 + \zeta_4^2 \\ \zeta_4 \end{bmatrix}. \end{aligned}$$

Now, in order to deduce the diffeomorphism $z = \phi(\zeta)$ to transform the regularized system into the output derivative injection observer normal form, we need firstly check Assumption 9.2, which yields $r_1 = r_3 = 1$ and $r_2 = 2$. So, we can calculate

$$\operatorname{rank}\begin{bmatrix} d\bar{h}_1 \\ d\bar{h}_2 \\ dL_{\bar{f}}\bar{h}_2 \\ d\bar{h}_3 \end{bmatrix} = \operatorname{rank}\begin{bmatrix} -1 & 0 & 1 & 1 \\ 0 & 0 & 1 & 2\zeta_4 \\ 0 & 1 & 2\zeta_3 & 0 \\ 0 & 0 & 0 & 1 \end{bmatrix} = 4,$$

which implies that Assumption 9.2 is fulfilled as well.

Finally, by following the procedure presented in Sect. 9.3, we obtain

$$\theta = \begin{bmatrix} d\bar{h}_1 \\ d\bar{h}_2 \\ dL_{\bar{f}}\bar{h}_2 \\ d\bar{h}_3 \end{bmatrix}$$

and the corresponding τ defined in (9.25). This enables us to obtain the following diffeomorphism

$$\begin{bmatrix} z_{1,1} \\ z_{2,1} \\ z_{2,2} \\ z_{3,1} \end{bmatrix} = \begin{bmatrix} \zeta_3 - \zeta_1 + \zeta_4 \\ \zeta_3 + \zeta_4^2 \\ \zeta_2 + \zeta_3^2 \\ \zeta_4 \end{bmatrix}.$$

With this deduced diffeomorphism, the studied nonlinear singular dynamical system might be transformed into the following form:

$$\begin{aligned} \dot{z}_{1,1} &= 0 \\ \dot{z}_{2,1} &= z_{2,2} + (2y_3 - 1)\dot{y}_3 \\ \dot{z}_{2,2} &= y_2 - y_3^2 + y_2 y_3 - y_3^3 + 2(y_2 - y_3^2)\dot{y}_2 + 4(y_3^2 - y_2 y_3)\dot{y}_3 \\ \dot{z}_{3,1} &= \dot{y}_3 \\ y_1 &= z_{1,1},\ y_2 = z_{2,1},\ y_3 = z_{3,1}. \end{aligned}$$

9.4 Observer Design

9.4.1 Nonlinear Luenberger-Like Observer

Consider the transformed normal form (9.4), this section will design a nonlinear Luenberger-like asymptotic observer for such a normal form. For this, we rewrite (9.4) into the following compact form:

$$\begin{aligned} \dot{z} &= A_O z + \alpha(\bar{y}) + \beta(\bar{y})\dot{\bar{y}} \\ \bar{y} &= C_O z, \end{aligned} \tag{9.29}$$

where

$$A_O = \mathrm{diag}[A_{O_1}, \ldots, A_{O_{\bar{m}}}]$$

$$\alpha = (\alpha_1^T, \ldots, \alpha_{\bar{m}}^T)^T$$

$$\beta = (\beta_1^T, \ldots, \beta_{\bar{m}}^T)^T$$

and

$$C_O = \mathrm{diag}[C_{O_1}^T, \ldots, C_{O_{\bar{m}}}^T]^T.$$

Due to the special form of A_O and C_O, it is easy to show that the pair (A_O, C_O) is observable. Therefore, we can always find a constant matrix G such that $(A_O - GC_O)$ is Hurwitz.

Before designing the nonlinear Luenberger-like observer, let us make the following assumption.

Assumption 9.3 It is assumed that $\beta(\bar{y})$ in (9.29) is integrable, i.e.,

$$\exists K(\bar{y}) \text{ such that } \frac{\partial K(\bar{y})}{\partial \bar{y}} = \beta(\bar{y}). \tag{9.30}$$

It is worth noting that the above condition for $\beta(\bar{y})$ (named as integrability) is a restrictive condition. For linear time-invariant systems where $\beta(\bar{y})$ in this case is constant (thus it is always integrable), the required integrability condition is satisfied. When considering nonlinear dynamical systems with single output, such an integrability is implicitly fulfilled, since in this case, we have $\bar{y} \in \mathbb{R}$. However, such a condition is not always fulfilled for nonlinear dynamical systems with multiple outputs, and that is why such a condition has been imposed when designing Luenberger-like observer for nonlinear dynamical systems with multiple outputs [7, 12].

Now, let us suppose that Assumption 9.3 is satisfied and consider the following dynamics:

$$\begin{aligned} \dot{\xi} &= N\xi + L(\bar{y}) \\ \hat{z} &= \xi + K(\bar{y}), \end{aligned} \tag{9.31}$$

where

$$\begin{aligned} &\tfrac{\partial K(\bar{y})}{\partial \bar{y}} = \beta(\bar{y}) \\ &N = A_O - GC_O \\ &L(\bar{y}) = G\bar{y} + NK(\bar{y}) + \alpha(\bar{y}). \end{aligned}$$

To show the convergence, let us define

$$e = z - \hat{z},$$

then we have the following observation error:

$$\dot{e} = A_O z + \alpha(\bar{y}) + \beta(\bar{y})\dot{\bar{y}} - N\hat{z} + NK(\bar{y}) - L(\bar{y}) - \tfrac{\partial K(\bar{y})}{\partial \bar{y}}\dot{\bar{y}}.$$

As it is assumed that Assumption 9.1 is fulfilled, then we have $\frac{\partial K(\bar{y})}{\partial \bar{y}} = \beta(\bar{y})$, with which the observation error reads as

$$\dot{e} = (A_O - GC_O)e + \alpha(\bar{y}) + G\bar{y} - GC\hat{z} + A_O\hat{z} - N\hat{z} + NK(\bar{y}) - L(\bar{y}).$$

In order to simplify the above equation, let $N = A_O - GC_O$, and

$$L(\bar{y}) = \alpha(\bar{y}) + G\bar{y} + NK(\bar{y}).$$

Hence, the above equation is equivalent to

$$\dot{e} = (A_O - GC_O)e,$$

where the matrix G can be freely chosen. In order to have the convergence, it is sufficient to choose a G such that $(A_O - GC_O)$ is Hurwitz. Consequently, we proved that (9.31) is an exponential observer of (9.29).

Based on the above result, we can state that $T\phi^{-1}(\hat{z})$ exponentially converges to x. In conclusion, the following dynamics

$$\begin{aligned} \dot{\xi} &= N\xi + L(y) \\ \hat{z} &= \xi + K(y) \\ \hat{\zeta} &= \phi^{-1}(\hat{z}) \\ \hat{x} &= T\hat{\zeta} \end{aligned} \tag{9.32}$$

is an exponentially nonlinear Luenberger-like observer of (9.29).

9.4.2 Design Procedure

Finally, for a given nonlinear singular system (9.1), the procedure to design a Luenberger-like observer can be summarized as follows:

Step 1: Calculate the elementary matrices T and S to transform (9.1) to semi-explicit form (9.12);

Step 2: Regularize (9.12) into a nonlinear regular system (9.15), if Assumption 9.1 is satisfied;

Step 3: If Assumption 9.2 and all conditions of Theorem 9.1 are satisfied, then deduce a diffeomorphism $z = \phi(\zeta)$ to transform the regularized system (9.15) into observer normal form (9.29);

Step 4: Finally, if Assumption 9.3 is satisfied, then construct the proposed Luenberger-like observer (9.32) to estimate x of (9.1).

References

1. Bejarano, F.J., Perruquetti, W., Floquet, T., Zheng, G.: State reconstruction of nonlinear differential-algebraic systems with unknown inputs. In: Proceedings of the 51st IEEE Conference on Decision and Control, pp. 5882–5887 (2012)
2. Bejarano, F.J., Floquet, T., Perruquetti, W., Zheng, G.: Observability and detectability of singular linear systems with unknown inputs. Automatica **49**(3), 793–800 (2013)
3. Bejarano, F.J., Perruquetti, W., Floquet, T., Zheng, G.: Observation of nonlinear differential-algebraic systems with unknown inputs. IEEE Trans. Autom. Control **60**(7), 1957–1962 (2015)
4. Boutat, D., Zheng, G., Boutat-Baddas, L., Darouach, M.: Observers design for a class of nonlinear singular systems. In: Proceedings of the 51st IEEE Conference on Decision and Control, pp. 7407–7412 (2012)
5. Campbell, S.L.V.: Singular Systems of Differential Equations. Pitman, London (1980)
6. Dai, L.: Singular Control Systems. Lecture Notes in Control and Information Sciences, vol. 118. Springer, New York (1989)
7. Darouach, M., Boutat-Baddas, L.: Observers for a class of nonlinear singular systems. IEEE Trans. Autom. Control **53**(11), 2627–2633 (2008)
8. Darouach, M., Zasadzinski, M., Hayar, M.: Reduced-order observer design for descriptor systems with unknown inputs. IEEE Trans. Autom. Control **41**(7), 1068–1072 (1996)
9. Paraskevopoulos, P., Koumboulis, F.: Observers for singular systems. IEEE Trans. Autom. Control **37**(8), 1211–1215 (1992)
10. Yip, E., Sincovec, R.: Solvability, controllability, and observability of continuous descriptor systems. IEEE Trans. Autom. Control **26**(3), 702–707 (1981)
11. Zheng, G., Boutat, D., Wang, H.: Observer design for a class of nonlinear singular systems. In: Proceedings of the 55th IEEE Conference on Decision and Control, pp. 3910–3914 (2016)
12. Zheng, G., Boutat, D., Wang, H.: A nonlinear Luenberger-like observer for nonlinear singular systems. Automatica **86**, 11–17 (2017)

Index

D. Boutat and G. Zheng, *Observer Design for Nonlinear Dynamical Systems*, Lecture Notes in Control and Information Sciences 487,
https://doi.org/10.1007/978-3-030-73742-9

N

O

P

R

S

T

Lightning Source UK Ltd.
Milton Keynes UK
UKHW020816080722
405567UK00002B/5

9 783030 737443